Edwaldo Henrique Bazana Barbosa
Arnaldo Yoso Sakamoto
Vitor Matheus Bacani

Environmental zoning for transboundary basins

Edwaldo Henrique Bazana Barbosa
Arnaldo Yoso Sakamoto
Vitor Matheus Bacani

Environmental zoning for transboundary basins

The Apa River Basin model

ScienciaScripts

Imprint

Cover image: www.ingimage.com

This book is a translation from the original published under ISBN 978-613-9-62943-5.

Publisher:
Sciencia Scripts
is a trademark of
Dodo Books Indian Ocean Ltd. and OmniScriptum S.R.L publishing group

120 High Road, East Finchley, London, N2 9ED, United Kingdom
Str. Armeneasca 28/1, office 1, Chisinau MD-2012, Republic of Moldova, Europe
Printed at: see last page
ISBN: 978-620-7-75833-3

SUMMARY

To the women in my life: Maria Catarina Barbosa (grandma), Izabel Cristina Bazana Barbosa (sis) and my daughter... goddaughter and sister Izabela Catarina Bazana Barbosa. I dedicate this book to you and thank you for the support you have given me during my journey of studies, professional and personal training and, above all, spiritual development.

ACKNOWLEDGMENTS

I would like to thank the Great Spirit for his protection and all those who directly and indirectly helped me to reach this scientific milestone, especially:

To Prof. Dr. Arnaldo Yoso Sakamoto, for his guidance, friendship, companionship, trust, financial support in times of crisis, for encouraging reading and the search for new theoretical foundations in the field of Physical Geography, I will always be grateful;

To Prof. Dr. Vitor Matheus Bacani, for his guidance, for his long-standing friendship, for his encouragement and support, for his patience during this time, for teaching me Geotechnologies and Physical Geography during my undergraduate studies, I will always be grateful for the friendship and trust placed in me over the years;

To Professors Dr. Ailton Luchiari (*in memorian)* and Dr. Wallace de Oliveira for their valuable contributions in the Qualifying Exam (master's degree), thank you for your suggestions, criticism and support;

I would like to thank Alice Yubero, from the *Centro de Estudios y Formación para el Ecodesarrollo - ALTER VIDA*, for providing official data from the Paraguayan government;

To my great love Keila Roberta Ferreira de Oliveira, for her friendship, companionship, encouragement, affection, love and for reaching out to me every step of the way, thank you for your support in putting this book together.

To my great comrades and friends, Kleber Rodrigo Penteado, Hermiliano Felipe Decco, Cesar Cardoso Ferreira and Leandro Félix da Silva for their great friendship, for their support in difficult times, for their partnership, for the moments of relaxation and for always being willing to help, thank you my great friends brothers, I will always be grateful for everything.

Epigraph

"People can die, but their ideas never will" (Ernesto Guevara de la Serna)

SUMMARY

The aim of this book is to present the reader with a methodological proposal for environmental zoning in transboundary basins, based on a physical-conservation diagnosis and prognosis of the transboundary hydrographic basin of the Apa River, using remote sensing data and geoprocessing techniques. The theoretical-methodological basis was the integrated analysis of the landscape from a geosystemic perspective. The methods included the correction and delimitation of the basin area, parametric evaluation of the terrain, temporal evaluation of land use and vegetation cover from 1987 to 2012, compilation of pedological and geological maps, application of environmental legislation and analysis of environmental fragility. The temporal mapping of land use and vegetation cover indicates that the main economic activity in the Apa River basin is livestock farming and extensive agriculture, which together account for 51.71% of the basin's total area. Environmental zoning was drawn up based on an environmental forecast of the transboundary Apa River basin, resulting in eight mapped environmental zones: a) Legal Restriction Zone (ZRL); b) Restricted Occupation Zone (ZOR); c) Conservation Units Zone (ZUC); d) Rural Productive Zone (ZPR); e) Rural Settlement Zone (ZAR); f) Indigenous Lands Zone (ZTI); g) Urbanized Zone (ZU) and, h) Incongruence Zone (ZI). The results contributed to the establishment of an environmental diagnosis and physical-conservation prognosis, which culminated in the Environmental Zoning of the Apa River Transboundary Basin.

Keywords: environmental zoning, transboundary basin, weighted overlap.

1. INTRODUCTION

Currently, the appropriation of space as a result of the current mode of production means that new areas are exploited or have their use intensified through anthropic actions. In this context, the landscape's structure is unbalanced, generating impacts on its evolutionary dynamics.

In this way, the landscape must be analyzed through a genic and morphogenic rescue in order to later analyze the correlation between physical characteristics and interventionist aspects (anthropic actions). Therefore, the concept of landscape is comprehensive in terms of physical aspects correlating with the relationship between man and the environment.

Considering the totality of the landscape, represented by physical and biological elements and by social relations, it can be seen that concerns about environmental issues are increasingly in evidence, since the natural environment is the basis for survival and is the stage for human relations. Technological advances have led to a significant increase in the urban network and growth in agricultural production in order to meet society's consumption needs. Therefore, economic growth actions must be planned and the aggressive agents of irrational exploitation processes must be circumvented.

Nowadays, a conservationist and preservationist methodology is essential for analyzing and investigating the real situation of the landscape in a given geographical space.

In this way, the physical-conservation diagnosis is considered a territorial planning tool. The diagnosis begins with a survey of the physical environment of the landscape, seeking to relate the physical aspects to the occupation of the environment. Based on the spatial data and attributes of the space, the process of drawing up conservation guidelines for minimizing the impacts caused by the current mode of production begins. These guidelines are part of land use planning.

The hydrographic basin is a fundamental planning unit of the earth's surface, considered to be the main physiographic unit of the terrain, because its characteristics govern the entire surface flow of water. It is therefore an ideal area for integrated planning of natural resource management in the environment it defines.

According to Rodriguez et al. (2011), the first conceptualizations of environmental planning in river basins were based on the approach of water management according to different types of use.

Today, the watershed is considered to be a combination of environmental factors, and in recent years the idea of environmental planning and management of all geo-environmental components has been developed, in accordance with the legislative prerogative of Law No. 9.433/97 establishing the National Water Resources Policy and *Law No. 3.239/07 on Paraguayan Water Resources.*

The physical limits of a hydrographic basin may go beyond the political-administrative limits of the federal units. In this case, watershed management will have to involve hydropolitical issues.

Hydropolitics should cover the political-administrative issues of the states that have border and/or transboundary watercourses. The main existing conflicts lie in the difference between the managing states (within legal limits) in legislating the use and exploitation of water resources.

This study will take a methodological approach to the conservation, maintenance, recovery and preservation of natural resources in transboundary river basins. This diagnosis is considered necessary to underpin other methodologies applied to transboundary basins. In general terms, the physical aspects, the production and occupation of space and political-administrative issues will be addressed and correlated, with a view to the integrated management of a river basin united by two countries.

The transboundary hydrographic basin of the Apa River is the area of analysis in this work, located between parallels 22° 36' 51" S and 20° 57' 37" S and meridians 55° 52' 28" W and 57° 59' 38" W.

Therefore, the aim of this work was to carry out an environmental zoning of the transboundary hydrographic basin of the Apa River, using remote sensing data and geoprocessing techniques. The synthesis of the physical and conservation analyses is expressed through thematic cartography.

2. OBJECTIVES

2.1. GENERAL OBJECTIVE

The main objective of this research was to draw up an environmental zoning plan for the transboundary basin of the Apa River, using remote sensing data and geoprocessing techniques.

2.2. SPECIFIC OBJECTIVES

- Mapping and evaluating terrain parametry using Digital Terrain Models (DTM);
- Mapping the temporal evolution of land use and vegetation cover in 1987 and 2012;
- Making a map of environmental legislation;
- Draw up a map of potential natural fragility to erosion;
- Produce a map of environmental fragility;
- Proposing environmental zoning;
- Draw up an Environmental Prognosis.

3. THEORETICAL-METHODOLOGICAL FRAMEWORK

3.1. The geosystem and ecodynamic theory: a theoretical contribution to landscape analysis

The concept of landscape, for most people (popular sense concept), expresses feelings, shows everyday aspects, brings back memories, in other words, feelings and experiences of the culture, customs, symbol, image or icon of a place, with value and liveliness.

For Santos (2006), landscape is the set of forms that, at a given moment, express the legacies that represent the successive localized relationships between man and nature. According to the author, landscape is a set of real-concrete objects. In this sense, landscape is trans-temporal, bringing together past and present objects, a transversal construction.

The landscape is the portion perceptible to the observer, where the combination of visible and invisible facts and their appropriate interactions are inscribed (TRICART, 1977).

Salgueiro (2001) points out that landscape has been identified, at least since the 18th century, with the physiognomy of a given area, its visible expression.

For Ab'Sàber (2003), the landscape is always an inheritance of physiographic and biological processes, and the collective heritage of the peoples who have historically inherited it as the territory in which their communities operate.

According to Salgueiro (2001), with the development of biogeography and some aspects of geomorphology that are closer to ecology lies the conceptual basis for the concept of landscape science, which has an ecological and profoundly naturalistic character and is well represented in France and Spain, with reference to Bertrand and the Toulouse school, as well as Wieber and the Besançon school, which, in methodological terms, are close to systems theory. The author states that the Besançon school tries to reconcile what is objective and subjective in landscapes through a systemic model.

Thus, Crepani et. al (2001), using a systemic theory, formulates the analysis of a natural landscape unit, listing the need for knowledge of its genesis, physical constitution, shape and stage of evolution, as well as the type of vegetation cover that develops on it. This information is provided by Geology, Geomorphology, Pedology, Phytogeography and Climatology and needs to be integrated in order to get a true picture of the behavior of each unit as it is occupied.

The concept of Geosystem, created in 1963 by the Soviet V. B. Sotchava (Pissinati and

Archela, 2009), consists of a theoretical subsidy for landscape analysis. In addition, landscape is a category of analysis in Geography, taking into account physical and social interactions, starting from a systemic analysis, with the aim of managing the development processes of territories.

In the course of his studies on the landscape, one concept that Bertrand revived was that of the geosystem, the definition of which was based on the interconnection of flows of matter and energy between biotic and abiotic elements Passos (1997 apud Pissinati and Archela, 2009).

Thus, Bertrand (1972) states that a geosystem is the result of the combination of ecological potential (geomorphology, climate, hydrology), biological exploitation (vegetation, soil, fauna) and anthropic action, not necessarily presenting physiognomic homogeneity, but an essentially dynamic complex.

Sotchava (1977) states that the so-called anthropogenic landscapes are variable states of primitive natural geosystems, which can be referred to the sphere of study of the problem of landscape dynamics.

Thus, Sotchava (op.cit) considers that geosystems are natural systems, at local, regional or global level, in which the mineral substrate, soil, communities of living beings, water and air masses, particular to the various subdivisions of the earth's surface, are interconnected by flows of matter and energy, as a single whole.

Physical environmental systems or geosystems represent the spatial organization resulting from the interaction of the physical component elements of nature (climate, topography, rocks, water, vegetation, animals, soils), having spatial expression on the Earth's surface and representing an organization (system) composed of elements, functioning through the flows of energy and matter, dominant in an areal interaction. The combinations of mass and energy, in the broad environmental energy control, can create internal heterogeneity in the Geosystem, expressing itself in a landscape mosaic. Alongside the vertical flows of matter and energy, as a function of the various structural horizons of ecosystems, there are flows in the spatial dimension connecting the various internal landscape combinations of the Geosystem (CHRISTOFOLETTI, 1999).

Corrêa (2009) defines a geosystem as a natural complex that has a spatial (territorial) pattern resulting from its history, its functional autonomy and the role it plays in the context in which it is inserted. He also emphasizes that geosystems are dynamic systems, as they have a history (temporality), which needs to be recomposed as the history of the dynamics of physical landscapes.

The study of geosystems allied to Physical Geography is capable of resolving numerous doubts related to the dynamics of the landscape and its morphological changes, with or without anthropogenic inducement. The theoretical contribution of geosystems indicates mitigating and compensatory issues for the environment, supported by planning and measures for the development and reconstruction of territories.

The Ecodynamic Theory advocated by Tricart (1977) is interpreted by Bacani (2010) as an evolution of geosystemic theory, one of the main theories guiding research in Physical Geography, based on the concept of landscape.

Ecodynamic Units are based on the logical instrument of a system and focus on the mutual relationships between the various components of the dynamics and flows of energy/matter in the environment. As such, the concept of Ecodynamics is based on the evolutionary dynamics of the landscape, correlating the weathering actions taking place in the environment.

In general terms, it can be seen that integrated landscape analysis is the modulation of the environmental system, through the characterization of its origin and pedogenetic formation, correlating morphogenic factors with the social conditioning factors or agents that produce space. This arrangement provides an understanding of the evolutionary dynamics of the landscape and the changes induced by anthropic action.

Therefore, the watershed can be considered a broad system with complex functioning, which justifies adopting a systemic analysis. In this way, Steinke and Saito (2009) point out that the geosystem is the best methodology to approach the study of an environment such as river basins, considering the effects of the interaction between physical and anthropic elements.

According to Diakonov (2002 apud RODRiGUEZ et. al, 2011), the hydrographic basin is a natural geosystem, the drainage networks are formed by the dialectical interaction of all the natural components, in varying degrees of naturalness, that is, in varying degrees of anthropogenic modification and transformation.

3.2. Hydrographic basin as a territorial planning unit

The notion of a hydrographic basin naturally requires the existence of water dividers, headwaters or springs, main watercourses, tributaries, sub-tributaries, as well as a hierarchy of drainage channels and a distribution of predominant soils TUCCI (2004).

According to Coelho Neto (1994), a drainage basin or hydrographic basin is an area of the earth's surface that drains water, sediment and dissolved materials to a common outlet at a given point in the river channel. The boundary of a drainage basin is known as the drainage divide or watershed. The drainage of a region not only depends on rainfall, topography, vegetation and soil texture, but is also influenced by the lithology and structure of the rocks.

The watershed area is defined by its physical and natural characteristics, which determine the functioning of the entire flow of matter and energy along its drainage network.

In addition to the physical aspects that characterize the watershed area, anthropogenic actions are also taken into account, which over the years have induced and/or "sculpted" new scenarios. According to Beltrame (1994), anthropogenic factors have a marked influence on the state of conservation of the physical and natural environment in the basin area. However, it is considered that this influence, due to its importance, deserves specific and in-depth studies, which could be done through a socio-economic diagnosis of the basin.

According to Santos (2006), territorial configuration is given by the set of natural systems that exist in a given area and the additions that humans have superimposed on these natural systems. As history unfolds, the territorial configuration is given by the works of men. A territorial configuration is created that is increasingly the result of a historical production, replacing the natural environment with an entirely humanized nature.

Santos (1988) points out that when all places are affected, directly or indirectly, by the needs of the production process, selectivity and hierarchies of use are created, with active or passive competition between the various agents, leading to a reorganization of functions between the different parts of the territory. Each point in space then becomes important, effectively or potentially. Its importance derives from its own natural or social virtues, either pre-existing or acquired through selective intervention.

In this sense, Ross (1992) argues that physical-territorial planning must take into account the potential of resources and the weaknesses of natural environments.

According to Rodriguez et al. (2011), the first conceptualizations of environmental planning in river basins were based on the approach of water management according to different types of use. Following this view, the watershed came to be seen as a combination of environmental factors, and in recent years the idea of environmental planning and management of all the watershed's geo-environmental components has been developed.

Physical-territorial planning consists of the conception of planning as the activity of drawing up spatial planning, with the plan functioning as a set of guidelines to be followed and goals to be pursued (in terms of land use, controlling the expansion of production, the provision of permanent preservation areas, etc.). Adapted from Taylor (1998 apud SOUZA, 2004).

Regional planning, or in a broader sense, territorial planning, is an activity linked to public power, with a political interventionist character, in order to guarantee the balance of the natural environment and the right to the city.

Territorial planning, as discussed here, uses the hydrographic basin as an area of political and administrative intervention. Based on this principle, the analysis must be detailed on a coherent mapping scale. Thus, planning on a general scale cannot be conceived without knowledge of the physical territorial aspects of an area.

Environmental legislation is an instrument for the coercive implementation of territorial planning. Its precepts exercise preventive actions in the environment, aiming for socio-environmental balance.

The mapping of environmental legislation can be defined as the spatialization of legislative aspects relating to the environment that can be mapped (BACANI, 2010).

The environmental legislation map is characterized as one of the main tools for land use planning, since combined with the maps of land use, vegetation cover and environmental fragility, it makes it possible to establish homogeneous landscape zones, where recommendations are proposed regarding ways of sustainably exploiting the environment (BACANI, op.cit).

Territorial planning uses a variety of instruments to facilitate decision-making on mitigating agents that degrade the natural landscape. Physico-conservationist diagnostics is considered a planning tool because it is based on geographic knowledge, due to its analysis of the integration between the physical-natural environment and human activities, in this case with the hydrographic basin as the physiographic unit.

In the proposal presented for the formulation of the physical-conservation diagnosis, the physical-natural elements of the hydrographic basin are focused on first; This includes the relief (geographical delimitation of the hydrographic basin, hypsometry and slope), the spatial dynamics of radiation (slope orientation), pedological and geological characteristics, the formation of vegetation cover and its domains, the spatialization of rainfall behaviour and the natural potential for erosion. In a second stage of the diagnosis, aspects relating to land use and vegetation cover are taken into account (which involves both those elements derived from natural dynamics such as natural vegetation, as well as those related to anthropic action, such as secondary vegetation, agriculture, buildings and other applications of production techniques in space), in addition to environmental degradation and compliance with environmental legislation (notably Law no. 12.651, of May 25, 2012).[1]

Integrated studies of a given territory presuppose an understanding of the dynamics of how the natural environment functions with or without human intervention. Thus, the preparation of environmental zoning must start with the adoption of a working methodology based on an understanding of the characteristics and dynamics of the natural environment and the socio-economic environment (ROSS, 1994).

The comparison between field data and cartography makes it possible to establish a proposal for environmental zoning, with guidelines for territorial planning of the river basin.

In order to carry out environmental zoning, it is necessary to know the reality of the geographical space and the specificities of the elements of the physical environment and society (SILVA, 2007).

Silva and Rodriguez (2007) consider that zoning is a planning product that proposes a type of functional use and a set of environmental functions that a given space should be subject to when planning for various scenarios. These notions reflect a certain difference from the process of ecological planning or zoning, which mirrors ecosystem organization and articulates the relationship between living organisms and their natural surroundings.

According to Morais 1997 (apud DEAMO, 2008), knowledge of current land use is an important prerequisite for integrated watershed planning. When it is represented on a cartographic plan, it becomes an indispensable material for defining the degree of protection provided to the soil by the current vegetation cover, the degree of degradation of the original vegetation cover, the rational use of the land, as well as helping to define its suitability for agricultural use. It also provides information on the type of production in the basin area.

The parametric evaluation of terrain includes the following cartographic documents: hypsometry, clinography and slope orientation OLIVEIRA (1984).

---

1 Provides for the protection of native vegetation; amends Laws Nos. 6.938, of August 31, 1981, 9.393, of December 19, 1996, and 11.428, of December 22, 2006; repeals Laws Nos. 4.771, of September 15, 1965, and 7.754, of April 14, 1989, and Provisional Measure No. 2.166-67, of August 24, 2001; and makes other provisions.

Parametric evaluation uses numerical terrain models (NTMs) to represent the features of the earth's relief.

Mapping the orientation of slopes makes it possible to observe the layout of their slopes in relation to their orientation; this layout is the basis for analyzing the flow of solar energy in the area, i.e. it makes it possible to identify the areas that receive the greatest or least amount of heat and light within a hydrographic basin (MENDONÇA, 1999).

Identifying and analyzing the hydrographic basin's hypsometry makes it possible to observe the altimetric variation of the area's relief, an important fact when analyzing processes related to the dynamics of land use (MENDONÇA, op.cit).

The slope of the land influences the infiltration of rainwater and the speed of surface runoff. If the land is unprotected by vegetation, the speed of surface runoff increases, contributing to an increase in erosion processes and increasing the load of sediment and other materials in watercourses.

Vegetation cover is important for maintaining renewable natural resources. As well as playing an essential role in maintaining the water cycle, it increases the porosity and permeability of the soil through the action of the roots, reducing surface runoff, maintaining humidity and soil fertility due to the presence of organic matter. The protection given to the soil by native forest results in less soil loss and greater water retention capacity (BELTRAME, 1994).

The greater importance of rainfall intensity is easily seen when it is observed that a high annual rainfall, but distributed evenly over the entire period, has a much lower erosive power than a lower annual rainfall that is poured out in torrents at a specific time of the year (CREPANI et al., 2001).

The hydrographic basin is a fundamental planning unit of the earth's surface, considered to be the main physiographic unit of the terrain, because its characteristics govern the entire surface flow of water within it. It is therefore an ideal area for integrated planning of natural resource management in the environment it defines. The territorial planning of a watershed must be systemic, analyzing all the potential factors for the protection and degradation of the watershed TUCCI (2004).

3.3. Hydropolitics and legislation in transboundary river basins

Elhance (1999 apud BROCH, 2008) defines hydropolitics as the systematic study of issues involving conflicts and cooperation between states that have water resources that transcend their political and administrative boundaries.

The concept of hydropolitics should encompass the political and administrative issues of the states that have border and/or transboundary watercourses. The main existing conflicts lie in the difference between the managing states in legislating the use and exploitation of water resources.

In practice, there are no institutions or authorities to resolve the different interests and coordinate the sharing of transboundary water resources according to the degree and hydrological interdependence of the shared systems in the bordering countries UNDP (2006 apud BROCH, 2008).

Each state has a particular need for use in the transboundary river basin. Territories can have a large representation in the basin area and exploit little of the watercourse, or the other way around.

The management of transboundary waters will also involve political and administrative issues linked to international politics, bilateral relations between the two countries and specific public policies for border areas (SAE, 2013).

In general terms, Lopéz (2004) suggests that the management of transboundary basins should take into account the strengthening and clear delimitation of the regulations and competencies of the different levels that make up the institutional entities or political-administrative units for the management of the river basin. In this sense, land-use planning must take into account future economic needs and population growth, in accordance with the carrying capacity and sustainable technical feasibility of use and occupation in the river basin.

In addition to the issues involving particularities and diverse interests in the use and exploitation of water resources, the planning and management of water in each political-administrative unit differs, leading to management conflicts in the hydrographic basin, which is then considered the planning unit.

Observing the provisions of Article 1, item V, of Law No. 9.433/97, which establishes the National Water Resources Policy, and *Article 3, item "d" of Law No. 3.239/07 on Paraguayan Water Resources,* the legislative instruments presented are similar in adopting the hydrographic basin as the basic unit for planning and managing water resources.

In Brazil, the Federal Constitution (BRASIL, 1988) deals with water in the provisions referring to public domain (arts. 20 II and 26 I), legislative competence (arts. 22 IV and 24 VI), administrative competence (arts. 23 VI and VII), the establishment of the National

Water Resources Management System (art. 21 XIX) and the right to an ecologically balanced environment (art. 225).

With regard to border and transboundary bodies of water, article 20, items I and II of the Federal Constitution (BRAZIL, 1988), confers them on the Union. Thus, the following are assets of the Union: lakes, rivers, any streams of water that serve as boundaries with other countries, or extend into foreign territory or originate therefrom, as well as marginal land and river beaches.

In the case of transboundary bodies of water, *Law No. 3.239/07* provides:

Article 8 *- The management of water resources shared with other countries shall be governed and/or regulated by the international treaties, conventions and agreements approved and ratified by the National Congress and in force.*

On December 11, 2002, Mozâo No. 14, published in the Official Gazette on March 10, 2003, approved that the Ministry of Foreign Affairs begin discussions with the Paraguayan government to develop a work agenda to promote shared management of the Apa River Basin.

Pereira et al. (2008) state that Brazil and Paraguay signed the Cooperation Agreement for the Sustainable Development and Integrated Management of the Apa River Basin on September 11, 2006, after three years of Mozao No. 14, aimed at the sustainable development and integrated management of the resources of the Apa River Basin. According to the author, the measure will enable the two countries to take coordinated actions that guarantee the preservation and rational use of the area's resources.

When it comes to joint management and land-use planning in a transboundary basin, Lopéz (2004) adds the need to strengthen local participation, since local actors play a key role in implementing actions to rehabilitate, protect, conserve, produce and manage the basins.

Law No. 9.433/97 and *Law No. 3.239/07* guarantee what was proposed by Lopéz (2004), stating that the management of water resources must be decentralized and participatory, including the public authorities, users and the community within the planning unit to which these regulations apply.

Shared and participatory management could be formulated through the creation of a transboundary river basin committee or mixed bilateral commissions. According to SAE (2013), it will require effective integration and cooperation, in harmony, between the different levels and bodies of management, monitoring and information in the countries involved.

The river basin committees are the local bodies of the Water Resources Management System (SINGREH), created by Article 32 et seq. of the Water Law. It is through the committees that management is decentralized at the level of the river basins, and they are responsible for "promoting debate on issues related to water resources and articulating the actions of the intervening entities", among other duties.

According to SAE (2013), basin committees play an important role in the management of transboundary waters, as they have the potential to identify conflicts of use or management of water resources originating in or related to the neighboring country, as well as to signal uses that may impact water management on the other side of the border. This importance is not only characteristic of the committees to be set up in border or transboundary river basins (under the control of the Union), but also in state-dominated basins and sub-basins that are tributaries of these. Identifying the uses, potential conflicts and weaknesses of river management in border areas is fundamental for defining shared management policies or even for concluding agreements and treaties. In this sense, the formation and operation of river basin committees in these regions should be encouraged.

3.4. Geotechnologies applied to physical and conservation diagnostics

The environmental, economic, social and cultural changes that took place in the second half of the 20th century and at the beginning of the 21st century have led to advances in different areas of scientific knowledge.

According to Rosa and Leite (2006), Quantitative Geography emerged in the Anglo-Saxon school, adopting neopositivism as its philosophical basis. This new current applied mathematics to geographical studies, as it believed it would make geography more precise. The theory of systems and models, as well as the use of statistics, were the evaluative methods that marked Quantitative Geography.

Accompanying the processes of transformation and adaptation of the sciences to new realities, some transformations have been significant in Geographical Science. In addition to the precision of geographic information, we can highlight the use of technologies applied to the recognition of territory (ROSA and LEITE, op.cit).

Geoprocessing was an instrument used and improved by Quantitative Geography in the 1970s, and is one of the major contributions to expanding geographic knowledge. With computing and mathematics as the basis for implementing this new method of study, geography, in its various forms, can provide its studies with a precise tool for spatial studies (ROSA and LEITE, op.cit).

Geoprocessing, or geotechnologies, is the set of technologies used to collect and process spatial information (ROSA, 2005). Geotechnologies involve the use of a set of computational and methodological resources to fulfill their functions, including GIS and remote sensing (ALMEIDA, 2009).

Thus, the term geoprocessing uses *Global Positioning System* (GPS), remote sensing, GIS and digital cartography to acquire data and process geographic information, which has been increasingly influencing the areas of cartography, natural resource analysis, urban and regional planning, among others (CAMARA and MONTEIRO, 2001).

The term geoprocessing is a technology or a set of technologies aimed at collecting and processing spatial information, which makes the term very broad in scientific knowledge, and encompasses several sub-areas, such as: Cartography, Remote Sensing, Global Positioning Systems (GPS, GLONASS, GALILEO), Topography, among others, in the collection stage, and in the integrated spatial data processing stage, Geographic Information Systems (GISs) stand out BACANI (2010).

GISs are one of the main tools of geoprocessing, which involves the collection and processing of spatial information, as well as the development of new systems and applications (ROSA, 2005).

A GIS is a system made up of *hardware, software* and data (LANG and BLASCHKE, 2009).

In Martinelli's (2003) view, GISs are automated devices for acquiring, managing, analyzing and presenting georeferenced data that is of interest to the space that is the object of geographic study, monitored over time, as well as providing simulations of complex events and situations in reality.

There are different concepts of GIS, within their perspectives and purposes, the authors do not show a consensus in the definition of GIS. Thus, Gonzâlez et. al (2012) point out that there are three different approaches:

"one based on functionalities, on the ability to create digital models of reality and on the potential of the tools that this discipline provides; another understood as an extension of the database concept itself, emphasizing the differences in terms of their structure and the different possibilities for handling spatial information; and the third dealing with aspects that we could call Organizational" or of the institutions and people who manage the geographic component, where GIS are configured as powerful information systems to help

management and decision making".

Therefore, the definition to be used for GIS (*Geographic Information System)* is:

"a set of computational tools composed of equipment and programs that, by means of techniques, integrate data, people and institutions, in order to make it possible to collect, store, process, analyze and offer georeferenced information produced by means of available applications, aimed at greater ease, security and agility in human activities related to monitoring, planning and decision making related to geographic space" ROSA (2005).

3.4.1. Components of a GIS

According to Rosa (2005), the components of a GIS are: hardware, software, data, users and analysis methodologies or techniques.

Hardware - This is the physical component of the computer structure itself, with its interconnected physical elements (computer and its peripherals), which are necessary for running the software. It also provides storage space for programs and geographic data.

Software - A program, embedded in an operating system, whose purpose is to collect, store, process and analyze geographic data. The software basically includes four modules:

- Data collection, standardization, entry and validation;
- Data storage and retrieval;
- Analysis and generation of information;
- Exit and presentation of results.

Data - This is a collection of maps and associated information in digital form (vector or raster). The database can be made up of two elements - a spatial database describing the shape and position of elements on the Earth's surface and an attribute database describing the characteristics or qualities of these elements. In some systems, the spatial and attribute databases are distinct from each other, while in others they are integrated very closely into a single entity. Spatial or geographic data can take three different forms: point, line and area. Spatial data is assigned values or attributes of the geographical space. Vector features are most commonly represented in *shapefile* (SHP) files. The SHP format is one of the most widely used standards for storing vector features in GIS and can be manipulated in various *software packages.*

Users - Persons trained to manipulate and operate the GIS, responsible for designing, programming, using and servicing the GIS.

Methodology or analysis techniques - According to Bacani (2010), methodology is a procedure that, based on a defined objective, subjects data to specific treatment in order to obtain the desired results.

Geotechnologies have only recently emerged in geographic science and with technological advances this science is also undergoing rapid technological changes, guaranteeing accurate data at a low financial cost. As a result, geotechnology is an important tool for analyzing watersheds and other analyses relating to the transformation of the natural landscape.

According to Beltrame (1994), the analysis of river basins can be enhanced with the use of Geographic Information Systems (GIS), in order to improve decision-making processes that require spatial information. In this way, the real entities of the geographical space are represented (roads, hydrography, vegetation cover and other types of use and occupation).

By creating a high-quality georeferenced database, it is possible to manipulate alphanumeric, raster and vector information in order to guarantee precision in decision-making regarding the planning and management of the analyzed watershed.

4. METHODOLOGY

4.1. Bibliographic, cartographic and iconographic survey

The methodological procedures for formulating the physical and conservation diagnosis of the transboundary basin of the Apa River were based on the proposals of: Beltrame (1994), Oliveira (1984), Mendonça (1999), Crepani et al. (2001) and Ross (1994).

Data collection and consultation of references in specialized literature were the initial stages of the research proposal. The following data was collected: cartographic materials and iconographic materials (raster images).

Secondary data was used, in *shapefile* format, from the Upper Paraguay Basin Conservation Plan - PCBAP (1997), the Radambrasil Project (1982), CPRM (1999), *Proyecto para la Protección Ambiental y Desarrollo Sostenible del Sistema Acuifero Guarani - Región Oriental del Paraguay (*2007), Macrozoneamento (1985) and Projeto GeoMS (2012) and topographic maps, on a scale of 1:100.000 scale from the Directorate of Geographical Services (DSG).

The iconographic materials used were:

- LANDSAT-5 satellite image, TM (*Thematic Mapper*) sensor; orbit point: 224/73, 224/74, 225/73, 225/74, 226/73, 226/74; bands 3, 4 and 5.
- Satellite image IRS-P6 (RESOURCESAT-1), sensor LISS-III - (*Linear Imaging Self-Scanner*), orbit point: 320/093, 320/094, 321/093, 321/094, 322/093, 322/094; bands 3, 4 and 5.
- SRTM radar image, containing a numerical grid interpolated by kriging to 30m by Valeriano (2008).

4.2. Parametric Land Evaluation

The parametric evaluation of the terrain was applied on the basis of digital terrain models (DEM), according to the procedures described in Oliveira (1984). The products used to describe terrain parametry were: altitude (hypsometry), slope (clinography) and slope orientation.

The SRTM (*Shuttle Radar Topography Mission)* interferometric RADAR image, with 30 meters of spatial resolution, freely available on the INPE website (www.dpi.inpe.br/topodata), was submitted to the *Focal Statistics* command, which in this case is an average filter with 3x3 cell dimensions. This procedure was necessary to avoid systematic errors during the image acquisition process. The "filtered" SRTM is then sent to the *Fill* command, which removes the spurious depressions. Both procedures were carried out in the *Spatinal Analyst Tools* extension of *ArcGis 10.0®*.

After the treatment process, the SRTM was used to generate the level curves, with an equidistance of 30 meters, which helped in the delimitation of the basin area, in the correction of the drainage network obtained from the GeoMS Project (2012) and in the spatialization of the relief features.

Subsequently, the hypsometric map was made by slicing it into nine classes, in order to contribute to an understanding of geomorphological compartmentalization and land use dynamics.

The clinogram map was based on the proposal by DeBiase (1992), and the radar image was used as the base map to define the analysis area, which culminated in the determination of slope classes. The definition of slope classes respects a particular nature, taking into account the needs of each researcher. In this case, five classes were defined for a better analysis of the results, according to the proposal described in Ross (1994).

The slope orientation map was drawn up to represent the different azimuthal degrees of exposure of the slopes, divided into four analysis classes with the following variations: 315° to 45° (N), 45° to 135° (E), 135° to 225° (S), 225° to 315° (W), as proposed by Mendonça (1999).

4.3. Temporal assessment of land use changes

Land use represents the techniques used to produce and transform geographical space. The conceptual basis proposed by the IBGE (2006) focuses on the observation and synthesis of the whole and the particularities of land use, oriented according to the geographical distribution of land resources, their social appropriation and environmental transformations, as well as technical survey and mapping procedures.

To generate the land use and vegetation cover maps for 1987 and 2012, image mosaics from the following satellites were used: LANDSAT-5, sensor TM (*Thematic Mapper*), dated August 1987, with bands 3B, 4R and 5G being used for the color composition and IRS-P6 (RESOURCESAT-1), sensor LISS-III - (*Linear Imaging Self-Scanner*), dated July 2012, with bands 3B, 4R and 5G being used for the color composition. Control points were used to register the images, based on the mosaic of georeferenced images from the LANDSAT-5 satellite, sensor TM (*Thematic Mapper*), dated May 2006. All the satellite images used in the mapping were obtained free of charge from the INPE website (www.dgi.inpe.com.br/CDSR), in digital format.

The satellite image mosaics were made using the automatic equalization process with a cut line, using the Histogram *Match* algorithm in *ArcGis 10.1®.*

The Interpretation Key was drawn up to help with the photointerpretation of geospatial features for the analysis of the years 1987 and 2012, as shown in Table 1.

Table 1. Interpretation key for mapping land use and vegetation cover.

Thematic classes	Samples	Color/Tone	Texture/Shape
Growing Areas	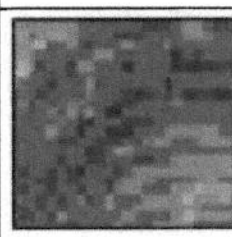	Blue/Cyan	Medium-rough Regular/Irregular
Water	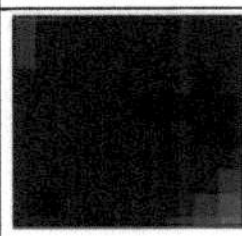	Dark blue	Smooth Irregular/Retilinear Curvilinear
Wet Areas	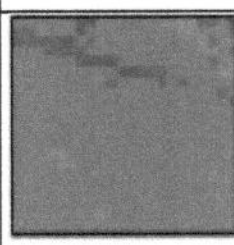	Moss Green Dark Grey	Medium-rough Irregular
Pasture		Green/Light Yellow	Medium-rough Regular/Irregular
Exposed Soil	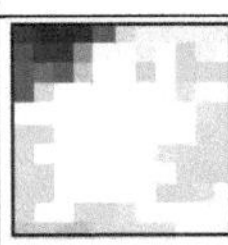	White	Smooth/ Irregular
Mata		Red	Rough/ Irregular
Reforestation	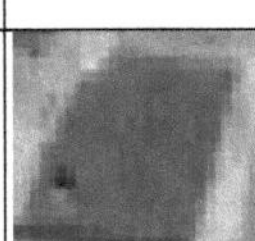	Red	Medium-Rough / Regular

Urban Area	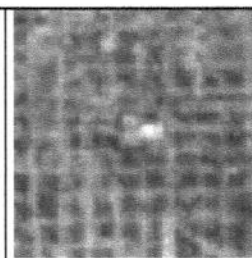	Mixture of colors, with cyan predominating, followed by magenta	Rough/ Irregular

Linear contrast was applied to all the images and they were then subjected to Supervised Regional Classification.

The algorithm used was Bhattacharya for the classification procedure. The Bhattacharya classifier is a supervised classification algorithm that requires the selection of training areas, and can use the regions separated during the segmentation process or polygons representing the regions to be classified. It uses the training samples to estimate the probability density function of the classes identified during training. It then evaluates the Bhattacharya distance between the classes in each region, according to Equation 1 (MOREIRA, 2005). The segmentation stage was based on the use of the region growing method. Using the exploratory method, the thresholds for similarity and area (*pixels)* were defined, corresponding to 5 and 20 respectively.

$$B(pi,pj) = \frac{1}{2}(m1-m2)^T \sum (mi-mj) + \frac{1}{2}\ln \frac{\left|\sum (mi-mj)\right|}{\left|\sum i\right|^{1/2}\left|\sum j\right|^{1/2}} \quad (1)$$

Where:

B = distance from Bhattacharya;

Pi and *pj* = pixels in classes i and j; *mi* and *mj* = averages of classes i and j;

T = transposed matrix;

ln = Neperian logarithm;

i and *j* = classes within the context.

Validation of the classification was based on the comparison of 100 field samples (carried out on June 1, 2012), with the distribution of values between the classes on the classified thematic map, according to the statistical parameters: *Kappa* concordance index (Equation 2) and overall accuracy (Equation 3):

$$\hat{K} = \frac{N\sum_{i=1}^{k} x_{ii} - \sum_{i=1}^{k}(x_{i+} . x_{+i})}{N^2 - \sum_{i=1}^{k}(x_{i+} . x_{+i})} \quad (2)$$

$$Acurácia_{global} = \left(\frac{\sum_{i=1}^{k} x_{ii}}{N}\right) \quad (3)$$

Where:

$\hat{K}$: estimated *Kappa* value; *k* : number of lines;

k : number of observations in row *i and* column *i;*

x_{ii} : sum of the elements of the matrix on its main diagonal;

$\sum_{i=1}^{k} x_{ii}$ total sum of observations for the rows;

x_{i+} total sum of observations for the columns;

N : total number of observations.

After using the classification algorithm, the thematic product went through the process of evaluating the performance of the classifier, where data obtained from the field (GPS points and photo-descriptive analysis) was used to compare the mapping with the data seen in the field. Corrections were made using matrix editing.

4.4. Spatialization of environmental legislation

The determination of Permanent Preservation Areas was based on Silva's (2003) proposal, based on proximity analysis, also known as *buffer* operation or corridor analysis. This operation consists of generating two-dimensional geographic subdivisions in the form of bands, whose outer limits have a fixed distance *x* and whose inner limits are formed by the limits of the geographic expression under examination. The proximity analysis is carried out in simple form, with a single band being determined. In this band, the Permanent Preservation Areas (PPAs) are spatialized in relation to the type of land use.

4.5. Rainfall behavior

The spatial rainfall variability was based on the annual rainfall averages available between 1970 and 2007 for the Brazilian side and the annual averages calculated from the data presented in the bulletins of the *Unidad de Gestión de Riesgos (UGR) del Ministerio de Agricultura y Ganaderia (MAG) for* 2007 for the Paraguayan side.

In Brazil, rainfall data from 18 (eighteen) weather stations of the National Water Agency (ANA) were used (available at http://www.cpao.embrapa.br/clima/). The weather stations are located in the municipalities of Amambai, Antonio Joao, Bela Vista, Bonito, Caracol, Coronel Sapucaia, Dourados, Guia Lopes da Laguna, Laguna Carapa, Maracaju, Nioaque and Porto Murtinho.

Six (6) weather stations were used in Paraguay, with rainfall data obtained from the

"Unidad de Gestión de Riesgos" bulletin (PARAGUAY, 2007). The weather stations are located in the departments of *Amambay*, *Alto Paraguay, Concepcion, Central* and *Pte. Hayes.*

For the spatialization of rainfall, the interpolation method called Kriging was used, which consists of a geostatistical method that takes into account the spatial characteristics of autocorrelation of regionalized variables (LANDIM, 2000).

Based on the values obtained for rainfall intensities and their relationship with vulnerability to soil loss, the maximum values of vulnerability to soil loss (CREPANI et al., 2001) and the degrees of fragility defined by Ross (1994) will be adapted, obtaining the degree of fragility and the weight, according to Equation 4.

$$IP = PMA/DPC \tag{4}$$

Where:

IP = Rainfall Intensity

AMP = Average Annual Precipitation

DPC = Duration of the Rainy Period

Based on the values obtained for rainfall intensities and their relationship with vulnerability to soil loss, a methodological adaptation was made based on a simple rule of three between the maximum values of vulnerability to soil loss (Crepani et al., 2001) and the degrees of fragility defined by Ross (1994).

4.6. Environmental Fragility

The analysis of environmental fragility was guided by the methodological adaptation of the proposals of Ross (1994) and Crepani et al. (2001) with the addition of the Areas of Biological Importance, recognized by Decree no. 5092, of May 21, 2004 and instituted by Ordinance no. 126 of May 27, 2004 of the Ministry of the Environment (MMA) as "Priority Areas for the Conservation, Sustainable Use and Benefit Sharing of Brazilian Biodiversity".

Knowledge of the potential of natural resources involves surveying the soil, relief, climate, flora, in short, all the components of the geographical stratum.

The cartographic products used in the analysis of environmental fragility were: Soils, Rainfall Intensity, Clinography, Biological Importance and Land Use and Vegetation Cover for 2012.

To draw up the environmental fragility map, the vector data was first converted into a matrix structure (*Grid.* format) and then reclassified using the *Reclassify* tool.

The data was then submitted to the map combination and weighted overlay methods.

For the reclassification and weighted overlay procedures, the *Spatial Analyst Tools* extension in *ArcGIS 10®* was used.

As proposed by Ross (1994), weights were assigned to the clinical and soil maps, assigning values to the mapped classes. As proposed by Crepani et al. (2001), weights (values) were assigned to the rainfall intensity classes.

The relationships between the mapped classes and the weights are shown in Table 1 and Table 2.

Table 1. Relationship between relief units and degrees of fragility.

Degrees of fragility	Clinography	Weights
Very weak	up to 6%	1
Weak	from 6% to 12%	2
Average	from 12% to 20%	3
Strong	from 20% to 30%	4
Very Strong	above 30%	5

Source: ROSS (1994).

Table 2. Relationship between degrees of fragility and soils.

Degrees of fragility	Types of soil Macrozoning (1985)	Weights
Average	Argissolo Vermelho-Amarelo with a clay texture	3
Strong	Medium-textured red-yellow Argissolo	4
Average	Red-Yellow Argisol with medium and medium/clayey texture	3
Average	Eutrophic Red-Yellow Argisol with a clay texture	3
Very Strong	Eutrophic Red-Yellow Argisol with a gravelly clay texture	5
Strong	Eutrophic Red-Yellow Argisol with a clayey and medium gravelly/clayey gravelly texture	4
Average	Eutrophic Red-Yellow Argisol with a medium/clay texture	3
Average	Eutrophic Red-Yellow Argisol with medium/clayey and sandy/medium texture	3
Average	Eutrophic Red-Yellow Argisol with a very clayey texture	3

Average	Eutrophic Red-Yellow Argisol with very clayey and clayey texture	3
Average	Rhendzic Chernosol	3
Average	Gleissolo Timórfico	3
Average	Gleissolo Timórfico with medium/clayey texture	3
Very low	Acric Red Latosol with a clay texture	1
Low	Acric Red Latosol with clay and medium texture	2
Low	Acric Red Latosol with medium and clayey texture	2
Very low	Acric red latosol with a very clayey texture	1
Very low	Aluminoferric red latosol with a very clayey texture	1
Very low	Eutrophic red latosol with a loamy and very loamy texture	1
Very low	Eutrophic red latosol with a clay texture	1
Very high	Litholic neosol with gravelly sandy texture	5
Very high	Litholic neosol with a sandy and medium gravelly texture	5
Very high	Litholic neosol with indistinct texture	5
Very high	Litholic neosol with indiscriminate gravelly and non-gravelly texture	5
Very high	Regolithic neosol with gravelly sandy texture	5
Very high	Regolithic neosol with a very gravelly medium texture	5
Average	Nitric Planosol with sandy/clayey texture	3
Average	Nitric Planosol with sandy/medium and medium texture	3
Average	Plinthosol with sandy/clayey texture	3
Low	Vertisol with clay texture	2
Degrees fragility	**of *Paraguay (1995)***	
Average	*Aquic Eutrochrept with fine French texture*	3
Strong	*Aquic Udifluvent with sandy texture*	4
Low	*Calcic Argiudoll with fine Arcillaceous texture*	2
Low	*Humic hapludult with fine Arcilla texture*	2
Low	*Lithic Hapludoll with fine Arcillaceous texture*	2
Low	*Lithic Udorthent with fine Arcillose texture*	2
Very low	*Mollic Paleudalf with a French-Greek texture*	1
Strong	*Typic Dystrochrept with sandy texture*	4

Low	*Typic Hapludert with fine Arcillaceous texture*	2
Low	*Typic hapludult with fine Arcillaceous texture*	2
Strong	*Typic Quartzipsamment with sandy texture*	4
Low	*Typic Rhodudult with fine Arcillaceous texture*	2
Strong	*Typic Udipsamment with sandy texture*	4
Strong	*Umbric Dystrochrept with sandy texture*	4
Low	*Vertic Paleudoll with fine Arcillaceous texture*	2

Source: Adapted from ROSS (1994) and *Paraguay (1995).*

The methodology used to revise the Priority Areas for the Conservation, Sustainable Use and Benefit-Sharing of Brazilian Biodiversity was discussed at the Workshop Updating the Priority Areas for the Conservation, Sustainable Use and Benefit-Sharing of Biodiversity - Targets and Tools, in November 2005 and subsequently approved by CONABIO Decision No. 39, of December 14, 2005. This methodology was based on the Map of Biomes of Brazil (IBGE, 2004) and used an approach that promotes greater objectivity and efficiency; creates a memory of the process of identifying priorities; promotes greater participation; and generates information that enables informed decision-making and the ability to assess opportunities (BRASIL, 2007). The areas were recognized by Ministerial Order No. 9 of January 23, 2007, in accordance with Decrees No. 2,519 of March 16, 1998 and No. 5,092 of May 21, 2004 (available at http://www.mma.gov.br/estruturas/sbf_chm_rbbio/).

The data on the areas of biological importance was first extracted from the Ministry of the Environment-MMA database (available at http://mapas.mma.gov.br/i3geo/datadownload.htm), in *shapefile* format, and then vector-edited to cut out the study area (*Arctoolbox/Analysis tolls/Extract*). The degrees of fragility were assigned taking into account the level of biological importance as shown in Table 2.

Table 2. Areas of Biological Importance and degrees of fragility.

Biological importance	**Degrees of fragility**	**Weights**
Extremely high	Very high	5
Very high	High	4
High	Average	3
Insufficiently known	Unknown/Null	0

Source: Probio/MMA.

The Ross (1994) methodology makes it possible to assign weights to the land use and vegetation cover map, assigning values to the mapped classes, as shown in Table 3.

Table 3. Land use and vegetation cover and degrees of fragility^

Thematic classes	Degrees of fragility	Weights
Growing Areas	High	4
Water	Very high	5
Wet Areas	Very high	5
Pasture	High	4
Exposed Soil	Very high	5
Mata	Low	2
Reforestation	Average	3
Urban Area	Very high	5

The resulting analysis maps are respectively: natural erosion potential (indicating potential fragility) and environmental fragility (indicating emerging fragility).

4.7. Environmental Zoning Proposal

Environmental zoning precedes environmental prognosis. This stage consists of analyzing the data obtained from the mapping of environmental fragility and environmental legislation.

The land-use mapping (zoning) of the Apa River transboundary basin consists of defining areas with different degrees of protection, with guidelines for the use and occupation of natural resources. To do this, the guidelines presented by the PCBAP (BRASIL, 1997) will be used, which considers the ideal economic production activity for the various types of soil formation, as well as determinations imposed by environmental legislation, notably the Forest Code (Law no. 12.651, of May 25, 2012), the degrees of fragility, Integrated Management of the Apa River Transboundary Basin (2003) and conservationist guidelines, among them Lepsch (2002), Mendonça (1997) and Beltrame (1994).

5. LOCATION AND CHARACTERIZATION OF THE STUDY AREA

5.1. Location of the Apa River transboundary basin

The transboundary hydrographic basin of the Apa River is located between parallels 22° 36' 51" S and 20° 57' 37" S and meridians 55° 52' 28" W and 57° 59' 38" W, as shown in Figure 1.

The capital of the state of Mato Grosso do Sul is approximately 326 km from the transboundary basin of the Apa River, and the main access routes are the BR 262, 419, 060 and 267 highways.

The area of the Apa River basin is approximately 15,433.00 km^2 , of which 11,481.8 km^2 are in Brazilian territory and 3,951.2 km^2 are in Paraguayan territory.

444 km long, the Apa River divides Brazil and Paraguay, with a drainage system that rises in the region called Cabeceira do Apa in the municipality of Ponta Porà, at an altitude of 708 meters (DSG, 1990). Its waters flow into the left bank of the Paraguay River, near the detachment of the Brazilian Army (2ª Cia de Fronteira), approximately 6.4 km from the mouth of the Apa River.

The municipalities that make up the Apa River basin in Brazilian territory, in the state of Mato Grosso do Sul, are: Ponta Porà, Antônio Joào, Bela Vista, Caracol, Porto Murtinho, Bonito and Jardim. In Paraguay, the basin area includes the Departments of Concepciòn and Amambay, and includes the following municipalities: Bella Vista, Concepciòn, Pedro Juan Caballero, San Carlos and San Làzaro.

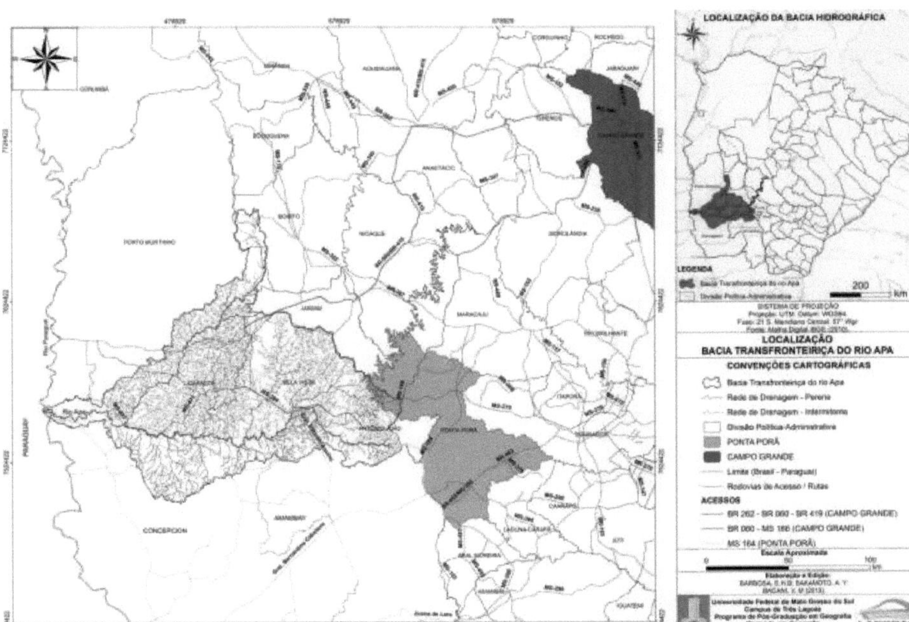

Figure 1: Location of the Apa River transboundary basin.

5.2. Physico-natural characterization of the Apa River transboundary basin

The Apa River basin is part of the Río de la Plata hydrographic basin, at the southern end of the Upper Paraguay river basin. According to CIDEMA (2003), the BAP stretches from the sources of the Paraguay River, in the region of Càceres, to the mouth of the Apa River. According to CIDEMA (op.cit), the Upper Paraguay river basin covers an area of 490,000 km^2 , of which 380,000 km^2 is in Brazilian territory and 111,000 km2 is on Paraguayan soil. This basin includes the Mato Grosso Pantanal, which is an extensive flood plain.

The cartographic base of the existing thematic mapping of the area in question is on a scale of 1:1,000,000, surveyed by the CPRM (1999) and *Proyecto para la Protección Ambiental y Desarrollo Sostenible del Sistema Acuifero Guarani - Región Oriental del Paraguay (*2007), PCBAP scale 1:250,000 (BRASIL, 1997) and scale 1:250,000 (MACROZONEAMENTO, 1985). The themes used to analyze the natural area of the Apa River transboundary basin are described below.

5.2.1. Geology

The geological framework of the Apa River basin is dominantly structured around the Apa Complex, formed by crystalline rocks in contact with the unconformity and tectonic types of rocks of the Jacadigo and Corumbà Groups, as well as being covered by Quaternary sediments of the Detrital Deposits, Xaraiés Formation and Pantanal Formation (BRASIL, 1982). The Apa Complex occupies an area of 4,263.67 km2, corresponding to 37.03 % of the total area corresponding to the Brazilian side of the Apa River basin.

In Paraguay, the region included in the Apa River basin comprises the *Alto del Apa* geological unit described in Paraguay (1986) as areas of exposure of crystalline rocks of Lower Precambrian and Upper Precambrian/Paleozoic age and their geological evolution affected the deposition of the Paranà River basin units.

Figure 2 shows the results of the geological mapping carried out by CPRM (1999).

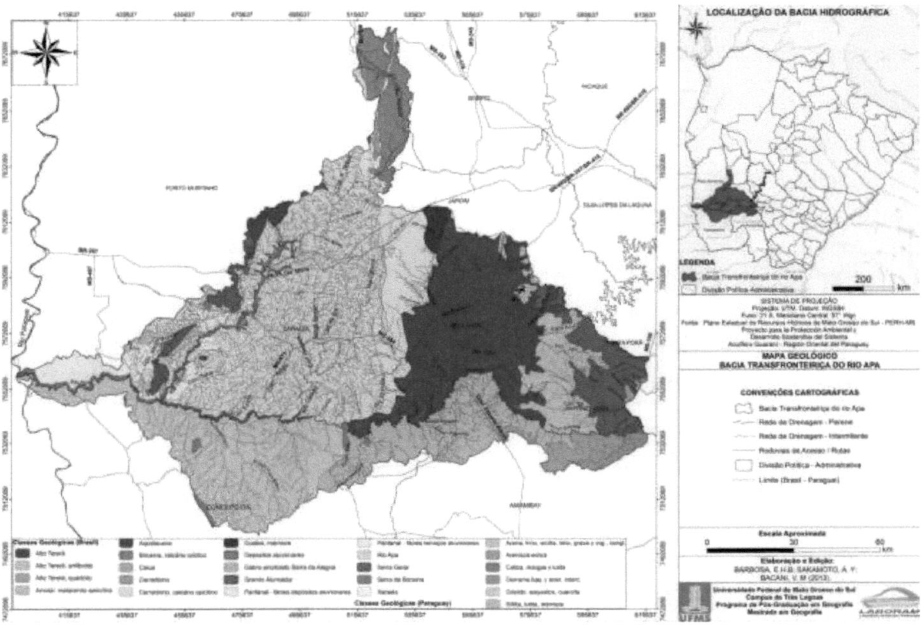

Figure 2 Geological map of the Apa River transboundary basin

The following geological classes are found on the Brazilian side of the transboundary Apa River basin:

a) Apa Complex: basically a group of crystalline rocks hidden under the Quaternary sediments of the Xaraiés Formation, Pantanal Formation and Detrital Deposits. Locally they can occur in small hills, generally on the lower slopes, overlying carbonate rocks from the Bocaina Formation or archosaurs and conglomerates from the Urucum Formation.

b) Aquidauana Formation: this unit is essentially made up of a sedimentary sequence with intense facies variation, predominantly sandy and purplish-red in color, in which three levels can be distinguished: Upper, formed by sandstones with cross-stratification and finely stratified brick-red siltstones; Middle, with fine to very fine sandstones, plane-parallel stratification and intercalations of siltstones, shales and diamictites, subordinate; Lower, containing reddish sandstones with diamictite lenses, intercalations of claystones, coarse whitish sandstones, arkose and basal conglomerate.

c) Cerradinho facies, calcitic limestone: Paraconglomerate, arkose, arkosian sandstone, sandstone with claystone laminae, shale with predominant intercalations of calcitic limestone. Coastal tidal plain environment with distal reworking of alluvial fans.

d) Faceis Amolar sericitic metarenite: Ortho-quarzitic metarenite, sometimes sericitic, with varied granulometry, found in coastal environments.

e) Serra Geral Formation: generally forms flat-topped reliefs and when the rocks are altered, they develop a characteristic red soil with dense vegetation cover, marked by the break in the relief. These are basaltic and basalt-andesite rocks of tholeiitic origin; rhyolite and rhyodacite. Sandstone, litharenite and volcanic sandstone interlayers.

f) Bocaina Formation, oolitic limestone: they are more resistant to erosion and generally form elevations of up to 600 meters in altitude, often standing out as residual hills in the Pantanal plain. They occur in marine platform environments subject to upwellings, formed by calcitic, dolomitic, intraclastic and oolitic rocks.

g) Pantanal Formation - Alluvial deposits facies: these are sedimentary formations that occur in the Paraguay River Depression, composed of alluvial sediments that are dominantly sandy, silty-clayey and clayey, unconsolidated and semi-consolidated. The sedimentation currently taking place in the interfluvial areas is conditioned by periodic flooding.

h) Alluvial deposits: these are unconsolidated alluvial sediments made up of pebbles, fine to coarse sands, with levels of gravel, lenses of silty-clayey material and remains of organic matter, related to flood plains, channel bars and current river channels.

i) Pantanal Formation - Alluvial terrace facies: these are sandy-clayey, semi-consolidated, partially laterized sediments.

j) Alumina Granite Body: Granite, porphyry microgranite, granophyry.

k) Alto Tererê Group: formed by garnet-mica schist, quartzite and lenses of amphibolite.

l) Cuiabà Group - Cuiabà Lithofacies: they appear in regions of flat topography, with tabular top relief forms. The rocks that make up this unit are represented in the area by schists, phyllites, metagrauvacas, metarchos, slates, metasiltites, marbles, quartzites, metaparaconglomerates and *homfels.*

m) Caiuà Group: made up of quartzose to subarcose, fine to medium sandstone, lenses of silty-clayey sandstone, sedimentary breccia, conglomeratic sandstone and conglomerate. Desertic continental environment: aeolian dunes, interdunes and ephemeral lakes.

n) Serra da Bocaina Suite: formed by rhyolite, dacite, rhyodacite and pyroclastic rock.

o) Cerradinho Formation: found in the northern part of the Bodoquena Plateau. Due to its lithological characteristics, this formation has been more eroded and forms flattened reliefs. The main feature is the lithological heterogeneity, marked by the alternation of sandstones, siltstones, shales, marls, limestones and dolomites, with the presence of arkose and locally conglomerates in the basal portion. It is conglomerate, arkose, sandstone, siltstone, claystone, marl, chlorite schist.

p) Alto Tererê lithofacies, amphibolite: composed of amphibole schist and amphibolite.

q) Xaraiès Formation: the origin of the deposits

Xaraiès is related to the opening of the Paraguay River Depression and the development of the Pleistocene pediplano, which took place under a semi-arid climate.

Almeida (1945 apud RADAM, 1982) distinguished four types of rock in this formation, with transitional passages between them: calcareous tuff with fossil veins; light, very spongy calcareous tuff; travertine with gasteropods; and conglomerate with calcareous cement.

r) Alto Tererê lithofacies, quartzite: formed by garnet-muscovite rocks.

s) Serra da Alegria gabbro-anorthosite body: made up of metagabbro, porphyritic metagabbro, metamelanogabbro, leucogabbro and metanorthosite formations.

Table 4 shows examples of the geological classes defined for the Paraguayan side of the Apa River transboundary basin.

Table 4. Geological classes mapped on the Paraguayan side of the Apa River transboundary basin.

Symbol	*Acuifers*	*Geological Period*	*Lithology*	*Area (km)²*	*Area (%)*
Pe	*Precambrian Shield*	*Precambrico*	*Granite, schist, cuarcite*	1832,43	46,76
Cco	*Colonel Oviedo*	*Carfonifero*	*Siltite, lutite, arenite*	1267,20	32,33
Qa	*Aluvion*	*Cuaternary*	*Arena, limo, arcilla, limo, grava y org., congl.*	372,43	9,50
Jm	*Misiones*	*Jurasico*	*Wind turbines*	280,98	7,17
Kb	*Basalt*	*Cretaceous*	*Bas. and aren. interc.*	99,82	2,55
Hey	*Itapucumi*	*Cambrico*	*Caliza, margas y lutita*	66,15	1,69
		Total		**3919,00**	**100,00**

Source: Adapted from *Proyecto para la Protección Ambiental y Desarrollo Sostenible del Sistema Acuifero Guarani - Región Oriental del Paraguay* (2007).

5.2.2. Geomorphology

The relief of the BAP is marked by significant contrasts between the periodically floodable lowlands of the Mato Grosso Pantanal and the surrounding, non-floodable lands, characterized by plateaus, mountain ranges and depressions. These large relief units are sculpted into various geological structures, which give them regional and local peculiarities, known as morphostructure (BRASIL, 1997).

On the Paraguayan side, the geomorphological units are related to the predominant lithological material in each unit. Thus, the following units are considered: *hills, mountain ranges, lomadas (high, medium and low), alluvial llanuras (high, medium and low or flooded) and alluvial valleys, with predominant types of lithology such as alkaline, sandstone, granite, basalt and sandstone* MOLINAS (2009).

Figure 3 describes the major geomorphological structures present on the Brazilian side of the Apa River transboundary basin, adapted from the PCBAP (BRASIL, 1997).

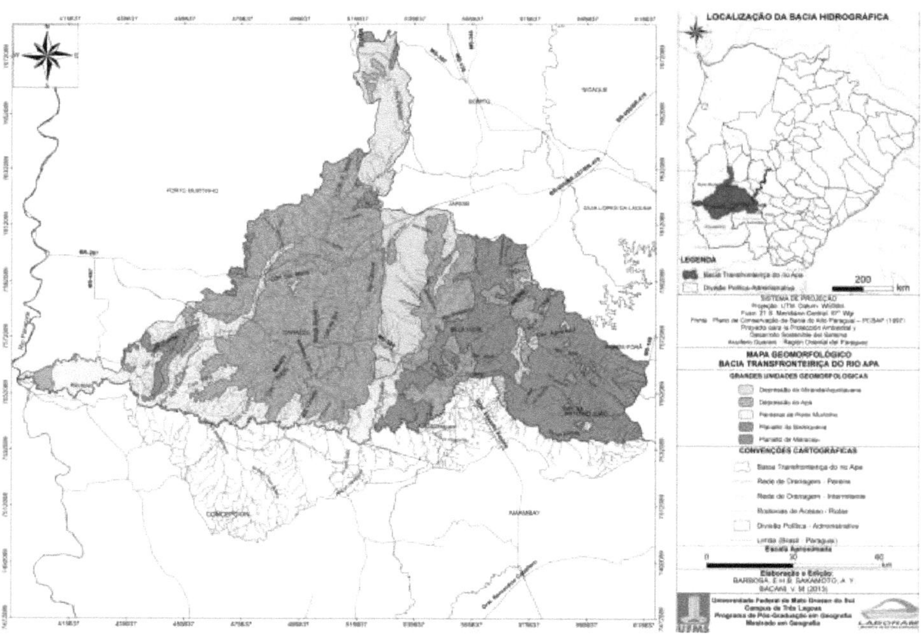

Figure 3 Geomorphological map of the Apa River transboundary basin on the Brazilian side.

5.2.3. Soils

According to Brasil (1997), in the Brazilian part of the Apa River transboundary basin, the most dominant classes are: Plinthosols, Nitric Planosols and Red-Yellow Argisols, according to an adaptation of the classification nomenclature proposed by EMBRAPA (2013). The land is used for natural pastures.

The soils mapped on the Brazilian side of the Apa River basin are described below.

a) Argissolo Vermelho-Amarelo: these are mineral, non-hydromorphic soils with a red to yellow textural B horizon and Fe2O3 contents normally lower than 11%. They have a sequence of horizons A, Bt, C or A, E, Bt, C, according to Camargo et al. (1987 apud BRASIL, 1997). Developed from very diverse lithologies, they vary greatly in their physical, chemical and morphological characteristics. The texture is generally sandy to medium and medium to clayey, in some cases with the presence of gravel. They are alkaline, dystrophic or eutrophic soils. They vary in relief from flat to undulating, although gently undulating and undulating predominate, and livestock farming is the main activity on these soils. The alkalic and dystrophic soils require liming and fertilization to minimize the toxic effect of aluminium and correct the low natural fertility, which makes them favourable for any agricultural enterprise. The terrain is not an obstacle to mechanization, but conservation practices must be carried out to minimize the erosive effects, due to its high susceptibility to erosion. They can therefore be used for any type of climatically suitable crop, however, given the great variability in their characteristics, it is necessary to consider some cases individually, with a view to their agricultural use.

b) Eutrophic Red-Yellow Argisol: occurring in areas of more rugged and dissected terrain, the texture is gravelly clay, with low-activity clay. On flat and gently undulating terrain, depending on the environment, the texture is medium/clayey, clayey and sandy/medium.

c) Rhendzic Chernosols: are mineral, non-hydromorphic soils, with a neutral or slightly alkaline reaction, from very shallow to shallow, consisting of a chernozemic A horizon above the limestone rock, or a C horizon derived from it, containing carbonate or even a calcic C horizon Camargo et al. (1987 apud BRASIL, 1997). According to Oliveira and Comar (2001), they occur with greater expression on the Bodoquena Plateau, on dissected surfaces in the form of small convex tops, resulting from the decomposition of limestone from the Bocaina Formation, i.e. on flat and gently undulating terrain, characterized by medium texture. Despite their high natural fertility, they are used for grazing colony grass, with the main limitations to more intensive use being the rock outcrops, the excess of gravel and pebbles and the strongly undulating and mountainous terrain when present.

d) Gleissolo Timórfico: these are mineral, hydromorphic soils with a glei horizon underlying an A horizon of moderate Camargo et al. (1987 apud BRASIL, 1997) or even weak, with a clayey texture. They are characteristic of flat, sloping sites subject to constant or periodic flooding. The anaerobic conditions, resulting from the poor drainage of the profile, give the soil intense gleaning characteristics, resulting from the reduction processes that intensify in these conditions. In general, the use of these soils is very difficult due to the conditions of poor drainage and periodic flooding, and they are more suitable for livestock farming.

e) Acric Red Latosol: these are mineral, non-hydromorphic soils, characterized by having a reddish red B horizon in shades 10R to 3.5YR, with iron oxide contents of between 8% and 18%, when clayey or very clayey, and normally less than 8% when of medium texture, with weak or no magnetic attraction (Camargo et al., 1987; Embrapa, 1988). They are normally very deep soils, rarely less than two meters thick, with high permeability and generally well to markedly drained. They have a sequence of A, Bw and C horizons, with a reduced increase in clay at depth. The main limitation to the agricultural use of these soils is their low natural fertility and the occurrence of toxic levels of exchangeable aluminium, which requires the application of correctives and fertilizers to make their exploitation viable. Nevertheless, the geographical conditions in which they occur, together with their great thickness, high permeability and lack of impediments to mechanization, give them excellent potential for intensive use. Because of this, the areas where these soils occur are practically devoid of their natural vegetation and are mainly used for soybean, corn and pasture crops.

f) Aluminoferric red latosol: found on flat terrain, with a very clayey texture.

g) Latossolo Vermelho Eutroférrico: varies according to the type of relief, in flat and gently undulating conditions, it has a clayey and very clayey texture, with high-activity clay.

h) Litholic neosols: these are poorly developed, shallow soils consisting of an A horizon directly above the rock or an incipient C or B horizon that is not very thick. In terms of possible uses, this class has a number of natural limitations, such as unfavorable topography for mechanization, shallow depth and in many cases the excessive presence of gravel and/or concretions, which act as an impediment to the development of roots, as well as chemical limitations in the case of alkalic and dystrophic soils. In view of the magnitude of the problems they present, the preservation of fauna and flora is the recommendation for these soils, although in some cases they can be used with restrictions for grazing and/or forestry.

i) Regolithic neosols: These are small isolated areas made up of quartzose sands, hydromorphic quartzose sands that occur within the Planosolos patches and alluvial soils that occur within the Plintossolos patches.

j) Nitric Planosols: occur to the south between the Apa and Paraguay Rivers and in the Serra da Bodoquena. These soils are characterized by the presence of a special type of textural B horizon, called B plânico, with a high clay content, underlying the A or E horizon, giving it an abrupt textural change. As a result, they have slow or very slow permeability, with a columnar or prismatic structure. This horizon incorporated the old nitric B horizon Embrapa (1999).

k) Plinthosol: There is a good percentage of individuals with alkalic and dystrophic characteristics.

l) Vertisols: These are soils made up of mineral materials, with a subsurface, verdant horizon, with peculiar pedological features due to the presence of friction surfaces (slickensides) due to the contraction and expansion of the clay, with cracks at least 1cm wide during the driest period of the year (Embrapa, 1999). It is located among rocks of the Apa Complex, near the city of Corumbà, and also includes some remnants of the Paraguay River Depression, but a large part of it is located in the Mato Grosso Plains and Pantanal near the Serra da Bodoquena.

The Paraguayan government, through the study carried out by Gorostiaga et al (1995), presents the definitions of the soils found:

a) *Aquic Eutrochrept:* These are soils rich in shifting bases and developed from calcareous sediments originating from sedimentary rocks. *Eutrochrept* generally have a thick vegetation cover and contain carbonates within the soil profile. It has a depth, measured from the surface of the mineral soil. It has a *fine loamy* texture.

b) *Aquic Udifluvent:* These soils are saturated with water at certain times of the year and have a sandy texture.

c) *Calcic Argiudoll:* this is a soil with a dark red color, with a mounded A horizon on a flat surface and an organic matter content of 3.5%, which indicates the cumulative effect of plant residues. It is dark greyish brown in color, sandy-clay and has a high calcium content. The pH is neutral or slightly alkaline.

d) *Humic hapludult:* Argissolo Vermelho is that they have a horizon DE imperturbable by no-till (AP) or of a horizon of 15 cm or thicker, that has a dark color, with a brightness in humidity of 3 or less and with brightness in drought of 5 or less.

e) *Lithic Hapludoll:* these have a clayey texture and occupy the areas of mountain ranges and isolated hills that are distributed along and near the Paraguay River; they form a cartographic unit characterized by material of calcareous origin, in steeply sloping areas with strong erosive surface drainage.

f) *Lithic Udorthent:* The soils classified as Udorthlient were evaluated for their dry matter production, with a good clay texture. They were recognized as covering mainly the hills of Caapucu, in the Department of Paraguari, from granite rocks, and in the hills of Santa Rosa as Missoes and the San Rafael mountain range, in Jtapua. All these landscapes are dominated by strongly undulating topography, with slopes of more than 25%, and surface drainage with high erosivity.

g) *Mollic Paleudalf:* The soils are Argissolo of different textural classes between 25 and 75 cm deep. They have been recognized as clayey, derived from sandstone and are found on gentle hills, occupying flat topographic positions to a lesser degree.

h) *Typic Dystrochrept:* Soils with modal Dystrochrept characteristics. They are soils with an ochre horizon and a cambic subsurface horizon, well-drained, deep and extremely acidic and with very low nutrient content. They do not have a lithic contact within 50 cm of the surface, the cambic does not contain a clay horizon, they are saturated with water for some time of the year, within 100 cm of the soil surface.

i) *Typic Hapludert:* These soils are dark in color at the top of the horizons, ranging from very dark grey to black, with a strong granular structure and subangular in very small forests that are friable when dry, and very plastic and very sticky when they contain more water. The clay texture is fine to very fine, which leads to low permeability. In depth, the horizons are slightly lighter in color, structured in small subangular blocks, and are also clayey, very plastic and sticky.

j) *Typic hapludult:* The surface horizons are dark yellowish brown to dark brown up to a depth of 40 cm. The soil texture is sandy loam, weakly structured, with cation exchange capacity.

k) *Typic Quartzipsamment:* these are deep soils, with no saturation conditions within 100 cm of the surface. The soils have a sandy Cambrian texture, well-drained soils with a similar morphology between horizons, with the only difference being the ochric horizon, since in the rest of the profile it presents a diffuse transition.

l) *Typic Rhodudult:* These soils are found in a mountain environment, where the source material is sandstone. They occur on flat topography, slope at a slight angle and are always well-drained.

m) *Typic Udipsamment:* The soils are sandy and deep, with a diffuse transition between horizons, which accumulates up to an organic matter content of 1.5%, with a sandy loam texture.

n) *Umbric Dystrochrept:* these are soils formed with Latosol, also sandy phase, derived from sandstone, distributed on hills, with slopes of less than 8%.

o) Vertic Paleudoll: These are mineral soils with a very clayey texture.

Figure 4 shows the soil classes present in the transboundary basin of the Apa River.

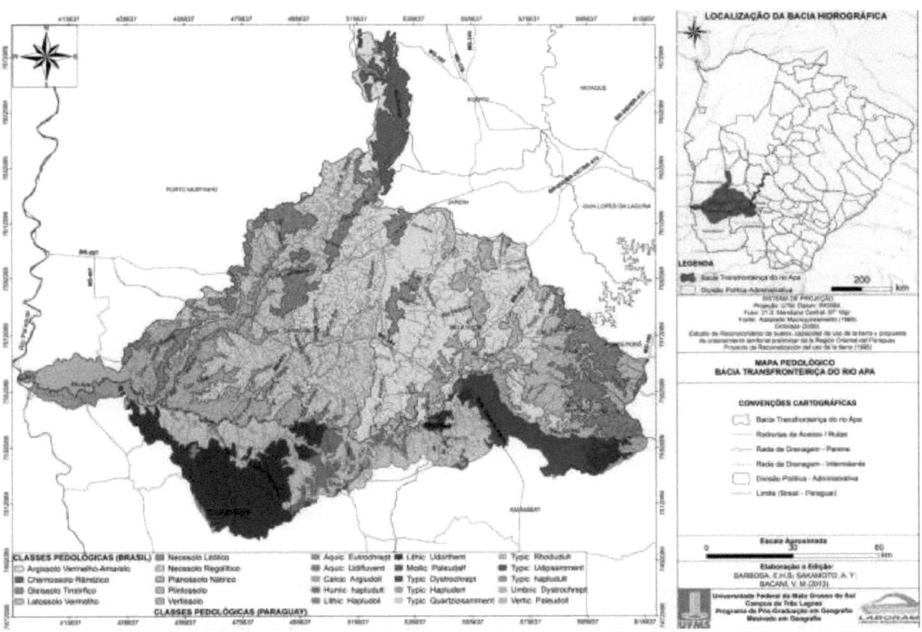

Figure 4 - Pedological map of the Brazilian side of the Apa River transboundary basin.

5.2.4. Vegetation

According to FADEMS (2012), the hydrographic basin lies within the Cerrado biome with great influence from the Chaco and the Atlantic Forest.

The vegetation cover of the Apa River basin is divided into: Dense Arboreal Savannah, Open Arboreal Savannah, Woody Grass Savannah and Steppe Savannah, according to SEMA (2005 apud BROCH, 2008).

In the transboundary basin of the Apa River, the following categories of analysis are considered: Cerrado (open woodland), Cerradoo (dense woodland), cultivated pasture (anthropic use), agricultural crops (anthropic use), gallery forest and hillside forest. As this is a transboundary basin, the classification on the Paraguayan side takes into account the *"natural communities of high importance for conservation," which are: Dense semi-deciduous subtropical forest (Bosque Alto), Dense semi-deciduous subtropical lowland forest (Bosque en galeria), Open semi-deciduous subtropical forest (Cerradón), Seasonally saturated subtropical grassland (Campo bajo) and Middle subtropical grassland with dry deciduous broadleaved trees (Cerrado)".*

Figure 5 shows the spatialization of vegetation domains according to MACROZENEAMENTO (1985), carried out only for the Brazilian side of the Apa River transboundary basin.

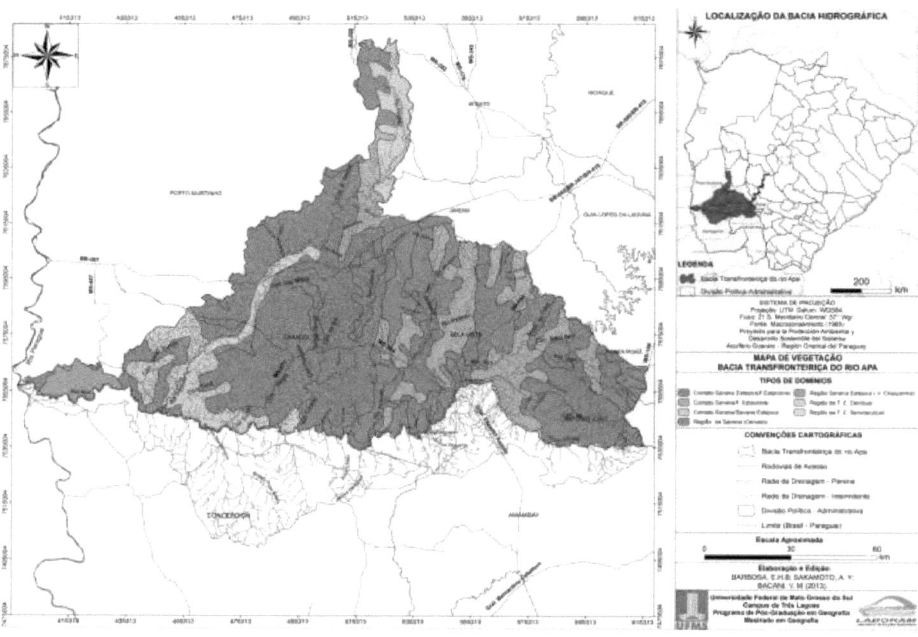

Figure 5 - Map of vegetation domains on the Brazilian side of the Apa River transboundary basin.

5.2.5. Weather

There are two climatic influences in the transboundary basin of the Apa River. According to Koppen's criteria (1948), the climate in the analyzed basin is of the *Aw* and *Cwa* types. The former is a tropical climate, with a dry winter, a rainy season in summer, from November to April, and a distinct dry season in winter, from May to October (July is the driest month). The average air temperature in the coldest month is over 18°C. Rainfall is over 750 mm per year, reaching 1,800 mm. The second has a subtropical climate, with a dry winter (temperatures below 18°C) and a hot summer (temperatures above 22°C).

According to Pereira et al. (2007), between the Serra de Bodoquena and the Serra de Maracaju - covering towns such as Guia Lopes da Laguna, Jardim, Bonito, Bela Vista and Antônio Joâo, at altitudes ranging from 250 to 550 meters, the climate is humid, with a water index of 20 to 40 and moderate water deficiency in winter. The temperature variation is small and the annual evapotranspiration is over 1,140mm.

Pereira et al. (op.cit) state that in the Pantanal Plain and almost all the depressions around it, the climate is dry sub-humid with a water index of -33.3 to 0.0. There is little or no water surplus in summer and water deficiency in winter. There is also little variation in temperature and annual evapotranspiration of over 1,140mm.

According to Llamas (1990 apud GOROSTIAGA, 1995), the Paraguayan side is defined as

The climate of this region is defined as Warm Subtropical towards the central part and Moderate Subtropical towards the extreme south, according to the methodology of the FAO's Agroecological Project for South and Central America. The average annual rainfall varies between 1,300 mm and 1,900 mm, and the average monthly temperature is between 17°C and 27°C.

Figure 6 shows the spatialization of rainfall in the transboundary basin of the Apa River.

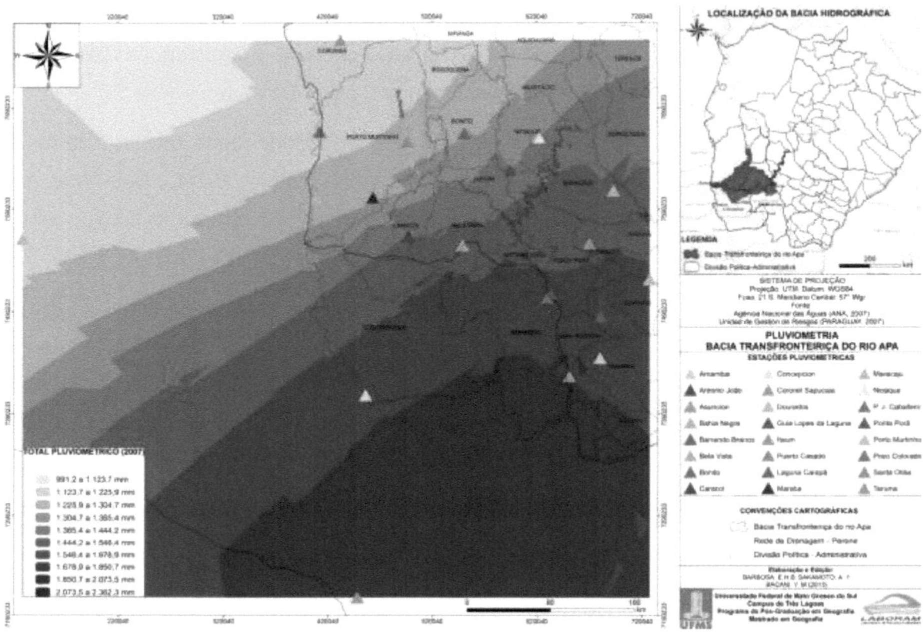

Figure 6 - Rainfall map of the Apa River transboundary basin.

5.3 Occupation history in the Apa River transboundary basin

Control of the territory in the south of Mato Grosso (now the state of Mato Grosso do Sul) was maintained through the slow and gradual possession of traditionally indigenous land, as presented by ESSELIN and OLIVEIRA (2007).

The position of the state of Mato Grosso in relation to the Spanish lands was one of the main reasons for the construction of several military garrisons on the banks of the rivers along the border in the second half of the 18th century. Thus, the southwestern border in the Paraguay River Valley received the most attention, due to its vulnerability and the fact that it was close to the Province of Paraguay ALMEIDA SERRA (1975 apud ESSELIN and OLIVEIRA, 2007).

For their part, the Castilians built the fort of San Carlos, on the banks of the Apa River, in 1777 ESSELIN and OLIVEIRA (2007).

Esselin and Oliveira (op.cit) point out that with the construction of the forts, the Spanish government began to distribute armored land to the settlers through concessions. The beneficiaries were relieved of the obligation to defend themselves, and the territory was made functional by the establishment of cattle and horse ranches between the Apa River channel and the middle Paraguay River (currently the Department of Concepción and the Department of Alto Paraguay in the neighbouring country).

The transboundary watershed region has been the scene of wars and disputes between the Guarani, *Payagua, Guaikuru* and Terena indigenous tribes (BROCH, 2008).

As an example of these wars and disputes, Esselin and Oliveira (2007) point out that the process of expropriation of land and enslavement of indigenous people, which began in the 16th century, intensified in the 19th century, especially after the war with Paraguay, due to the expansion of pastoralism undertaken by pioneers who settled in various parts of the province, especially in the Pantanal. As a result, the indigenous population was dispersed as cowboys and farmers.

The Indians were considered cheap labor on the farms that were being organized in the south of Mato Grosso. According to Esselin and Oliveira (2007), the main activities of the Indians were: harvesting and preparing ipeca grass, rubber, as well as being boatmen or oarsmen.

The *Guaikuru* Indians occupied the region from the south of Corumbà to the Apa River, encompassing the Pantanal region of Nabileque, for their festivals, burials and other ceremonies. According to the author, this was the heart of *Guaikuru* territory, and it was to this place that they always returned, where their villages themselves were located. The *Guaikuru* Indians were easily mobile, considered to be hunter-gatherers, always looking for new territories.

Another occupation of the Apa River region took place when the Jesuit priests abandoned the Itaim region in 1649. After fighting the bandeirante Raposo Tavares and his commanders, the Jesuits and the Indians who managed to escape crossed the Apa River and settled in Paraguay (ESSELIN and OLIVEIRA, 2008).

The *Guaikuru* Indians began looting and raiding various properties located in the Rio Apa region. Over time, the raids intensified and became increasingly daring.

Esselin and Oliveira (2007) report that around 3,000 head of cattle and 1,000 horses from the ranches located on the banks of the Apa River were attacked by the Indians. But in this attack, the Castilians, forewarned, managed to recover most of the animals, which were already a few leagues north of the Apa River.

Still according to Esselin and Oliveira (op.cit), the attacks on Castilian farms became more and more intense and ended up making it impossible for the Spanish to delimit their borders peacefully. The situation required the organization of troops to repel the Indians.

The priority was to garrison the southern entrance to the Apa River, through which they would be attacked.

Mello (1959 apud ESSELIN E OLIVEIRA, 2007) mentions that the Indians, using their equestrian skills and knowledge of the cow fields, continued to plunder, pillage and steal cattle.

In addition to repelling the Indians, the Castilians wanted to consolidate the territory south of the Apa River, considering that their attempts to continue north up the Paraguay River were prevented by the actions of the Portuguese garrisons.

Esselin and Oliveira (2007) say:

"the *Guaikurus* were docile friends to drive the territorial demarcation stakes, distancing the Spaniards, while the Castilians were docile enemies who were essential to minimize the strength of their friends."

Thus, war was declared, the Spanish invaded Portuguese territory armed and with the "help" of the Portuguese, captured and killed the *Guaikurus* who were considered allies of the Portuguese.

The Portuguese-Brazilians, as Esselin and Oliveira (2007) call them, were afraid that the "fury" of the *Guaikurus* might turn against them. By being complacent with the attacks of their Castilian rivals, the Portuguese ensured the weakening of the groups and the beginning of indigenous disintegration. And the previous attacks on the haciendas in the Apa River region weakened Castilian colonization.

Due to the raids by the *Guaikurus* Indians, there was an increase in loose, untamed cattle in the Apa River region. Esselin and Oliveira (2007) conclude that the cattle scattered throughout the Pantanal would attract more settlers (from São Paulo and Cuiabá) to increase the value of the area.

Esselin and Oliveira (2008) show that the main reason for the displacement of the Cuiabá people to the southern Pantanal, known as the vacarias, was the large number of farmers fleeing justice for crimes committed against the Portuguese in various cities in Mato Grosso, around Cuiabâ. So they settled along the Paraguay River, populating the banks of the Taboco and Nioaque Rivers, moving on to the valleys of the Miranda, Aquidauana and Negro Rivers, reaching the vicinity of the Apa River.

Esselin and Oliveira (2007) observe that in this context, borders have become demarcating instances of intersectoral relations of legitimization of power and recognition of sovereignty.

According to Esselin and Oliveira (2008), the colonization process of the Southern Pantanal (including the Apa River region) has been linked to cattle ranching since the beginning of the 16th century. The limitations imposed by the environment were overcome by cattle, which ensured the economic occupation of the plain and contributed decisively to the incorporation of the region into the national market.

In addition to livestock farming, according to Broch (2008), the economic cycle of yerba mate defined investments and socio-cultural relations in the region, along with the native fields, covered in "goat's beard", which served as natural pasture for extensive beef cattle breeding.

The historical factors of occupation of the territories contributed to the cultural formation of the populations of the Apa River basin region, completed by the existing miscegenation, from the arrival of the Spanish, Portuguese and migrants from various places in Brazil, especially the southern region MEDEIROS (2008 apud BROCH, 2008).

According to Broch (2008), the Guarani left a cultural mark on the people of the region, evidenced by the Guarani language itself, which is considered one of Paraguay's national languages and is spoken frequently by Paraguayans and Brazilians on the border with Brazil.

By understanding the process of occupation of the territories consolidated within the transboundary basin of the Apa River, the evolutionary form of land use and occupation can be better understood. Therefore, the peculiarities of the historical, cultural, social and economic aspects transform the environment as the factors or conditioning factors that shape space change.

6. Results and discussion

6.1. Parametric Land Evaluation

Parametric terrain evaluation quantifies the relief by presenting the following cartographic documents: hypsometry, clinography and slope orientation.

Contour lines were generated from a mosaic of 30m SRTM Radar images (TOPODATA), with an equidistance of 30m. The identification and hypsometric analysis of the Apa River basin made it possible to analyze the altimetric variation of the relief, delimit the basin area and extract the drainage network using vector editing.

Studies point to controversies surrounding the definition of the area of the Apa River basin. According to CIDEMA (2003), the Apa River basin in Brazil covers an area of 12,000 km^2. Prior to this delimitation and based on meetings of the Technical Chamber for the Management of Transboundary Water Resources - CTGRHT, the Apa River basin would have a physical area of 15,617.53 km2, of which 12,181.31 km2 would be in Brazilian territory and 3,436.22 km2 in Paraguayan territory.

Based on this research, the area of the Apa River basin is considered to be 15,432.72 km2, of which 11,513.72 km2 is in Brazilian territory and 3,919 km2 in Paraguayan territory.

Hypsometry shows us the topimetric knowledge of the area by means of elevation points and contour lines. The profiles show us the morphometric characteristics of the area (FERREIRA, 2011).

The hypsometric map, as shown in Figure 7, indicated a predominance of altitude variation from 204 to 257 meters, totaling an area of approximately 3,206.57 km2, making up 20.78 % of the total basin area. These areas are located in the central-western portion of the basin, expanding eastwards.

Table 5 describes the classes of altimetric compartments in the Apa River basin.

Table 5. Altimetric compartment classes.

Altimetric classes	Area (km)^2	Area (%)
68 to 148 meters	1.889,31	12,24
148 to 204 meters	3.110,44	20,15
204 to 257 meters	3.206,57	20,78
257 to 313 meters	2.513,75	16,29

313 to 375 meters	1.919,29	12,44
375 to 444 meters	1.222,49	7,92
444 to 521 meters	641,38	4,16
521 to 604 meters	627,03	4,06
604 to 747 meters	302,47	1,96
Total	**15.432,72**	**100**

The high altitudes comprise the 521 to 604 meters and 604 to 747 meters classes, located to the north in the Bodoquena mountain range and to the east at the source of the Apa River, in the region known as Cabeceira do Apa.

To the north of the Apa River basin, in the Serra da Bodoquena, the Serra da Bodoquena National Park was established by Federal Decree on September 21, 2000, with a total area of 76,826.13 ha, 37.31% of which is in the Apa River basin, comprising an area of 28,666.46 ha. This shows the degree of protection determined by law for this area.

The Apa River rises in the region known as Cabeceira do Apa in the municipality of Ponta Porâ, at an altitude of 708.00 meters (DSG, 1990). Its waters flow into the left bank of the Paraguay River, at an altitude of 80.00 m, near the Brazilian Army detachment (2ª Cia de Fronteira), approximately 6.4 km from the mouth of the Apa River.

Figure 7 shows the spatialization of the altimetry classes in the transboundary basin of the Apa River.

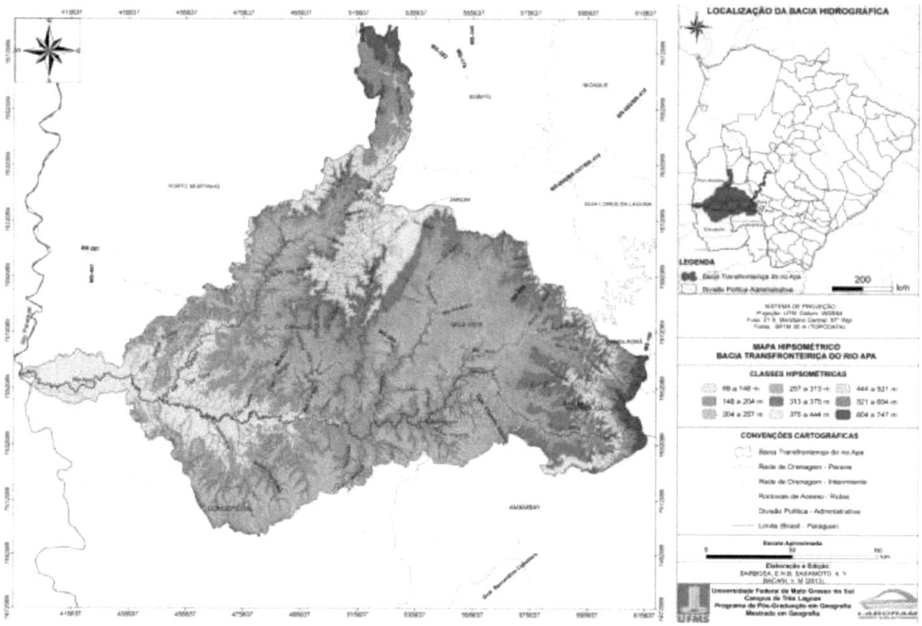

Figure 7 - Hypsometric map of the Apa River transboundary basin.

The length of the longitudinal profile of the Apa River is approximately 315 km, with altimetry varying from 550 m in the upper reaches to 80 m in the lower reaches, with a range of 470 m. Figure 8 shows the altimetric gradient.

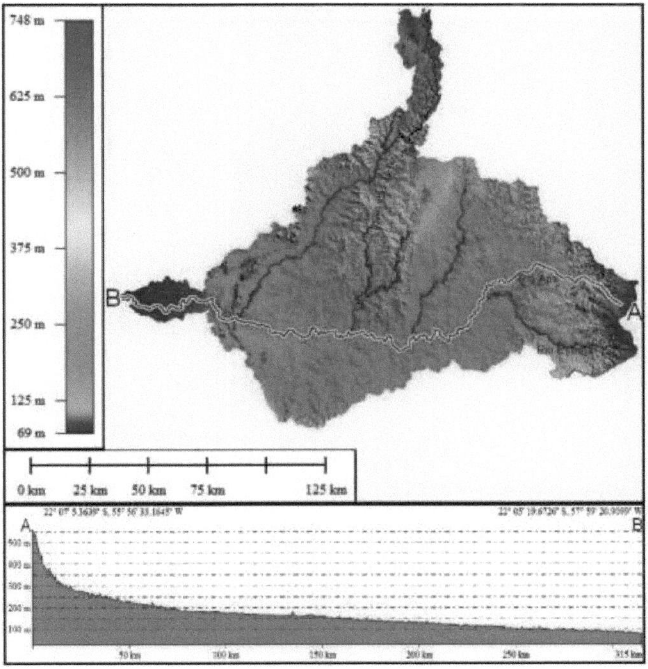

Figure 8. Longitudinal profile (section A - B) of the Apa River.

According to the classification of relief forms, based on the amplitude cited by Moreira and Neto (1998), the longitudinal profile shows a Dissected Mountain type relief.

Relating the longitudinal profile shown in Figure 7 to the geological structural base of the terrain, formed by sedimentary rocks, generated a Tabular relief form (flat horizontal bedding).

Another important watercourse is the Estrela River, which rises in the municipality of Antonio Joao at an altitude of 699 m and flows into the Apa River in the municipality of Bela Vista.

Figure 9 shows the cross-sectional profile near the source of the Apa River and the Estrela River, covering a distance of 37.7 km. In this profile, the altimetry varied from approximately 546 m to 545 m, with an amplitude of 20 m, typical of flat terrain, according to Moreira and Neto (1998).

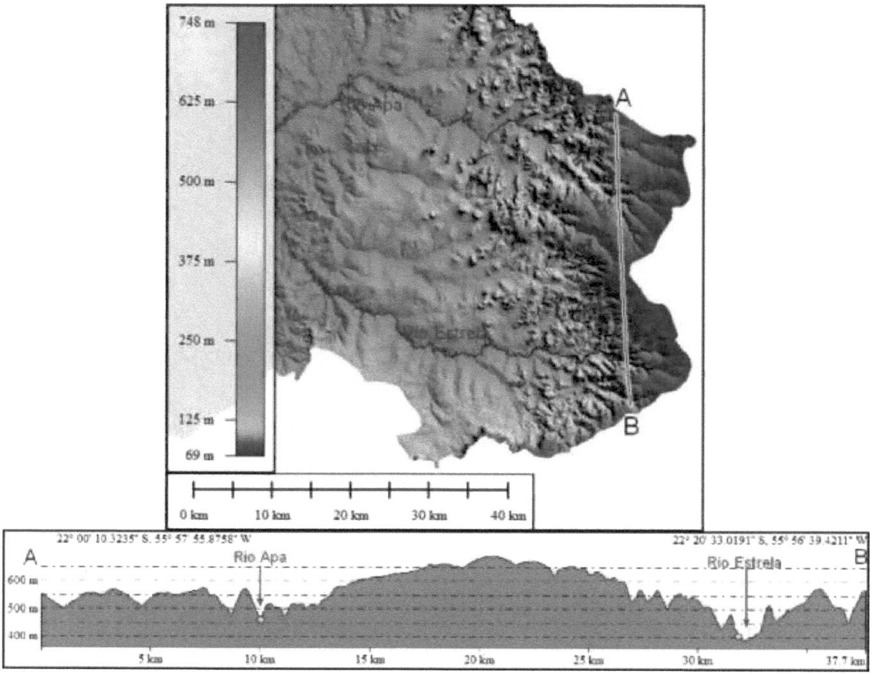

Figure 9. Cross-sectional profile (section A - B) near the headwaters of the Apa River and Estrela River.

In the middle course, the cross-sectional profile shows a distance of 171 km. The altimetry varies from 600 m to 350 m, with a range of 250 m. According to Moreira and Neto (1998), the relief is of the Dissected Hill type. Figure 10 shows the profile of the middle course of the Apa River basin.

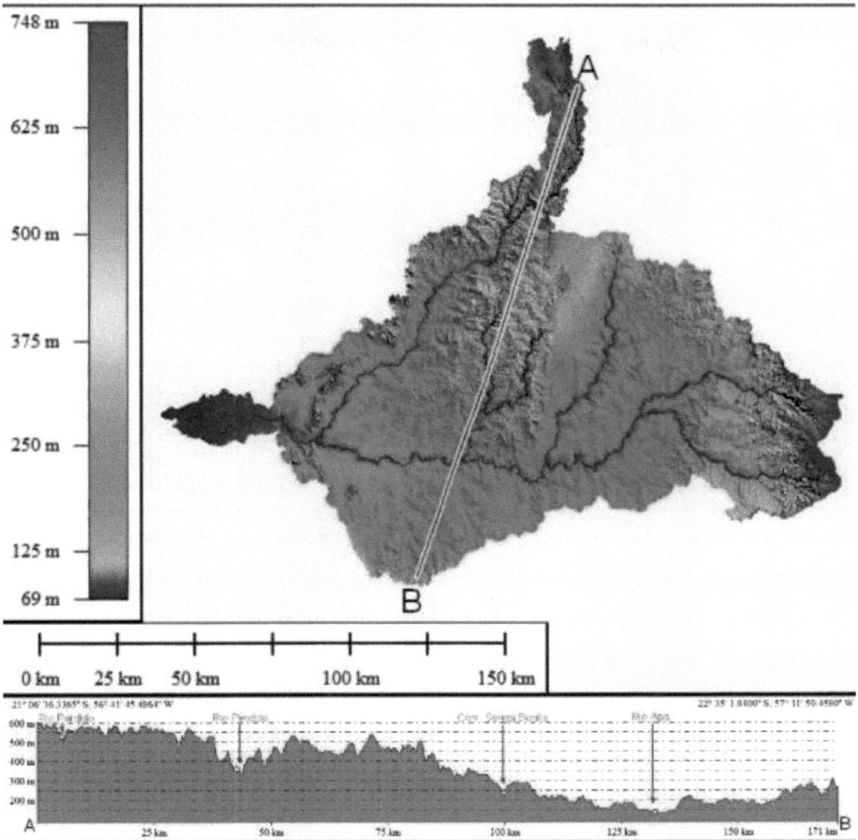

Figure 10. Cross-sectional profile (section A - B) in the middle reaches of the Apa River basin.

Figure 11 shows the cross-sectional profile drawn near the lower course of the Apa River basin at a distance of 12.21 km. In this profile the altimetry varied from 85 m to 115 m, with an amplitude of 30 m.

The relief shown by the longitudinal profile is flat, according to the classification of Moreira and Neto (1998).

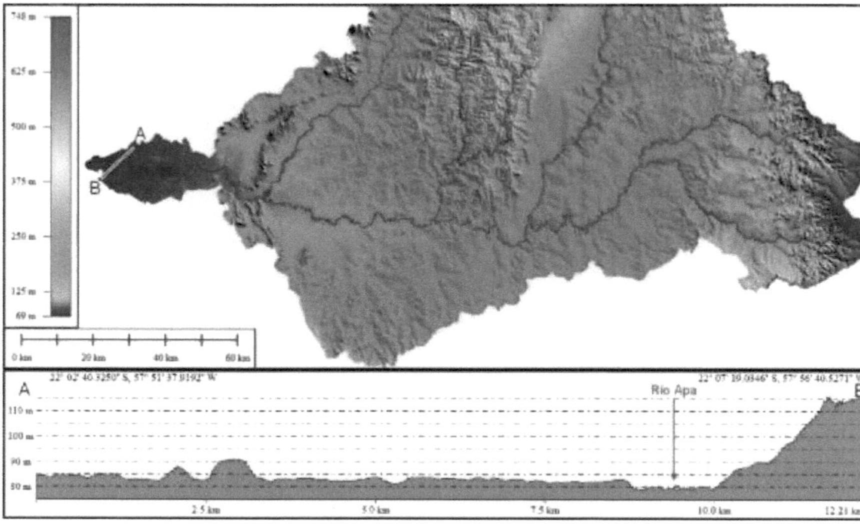

Figure 11. Cross-sectional profile (section A - B) in the lower course of the Apa River basin.

Clinography mapping (Figure 12) shows that the steepest areas are at the springs, where slopes of more than 30% were found, which tends to increase the flow of surface runoff and the dragging of matter from the rocks and soil. According to Mendonça (1999), in areas with steeper slopes, landslides occur, resulting in agricultural losses. The erosive effect will be enhanced by the removal of the vegetation cover near the watercourse, contributing to the siltation process. The removal of the vegetation cover alters the procedural relationships, so that rainfall is no longer intercepted, leading to the mechanical disintegration of the soil through the *"splash"* effect, while at the same time causing an increase in overland flow with consequent desilting, gullies or even landslides.

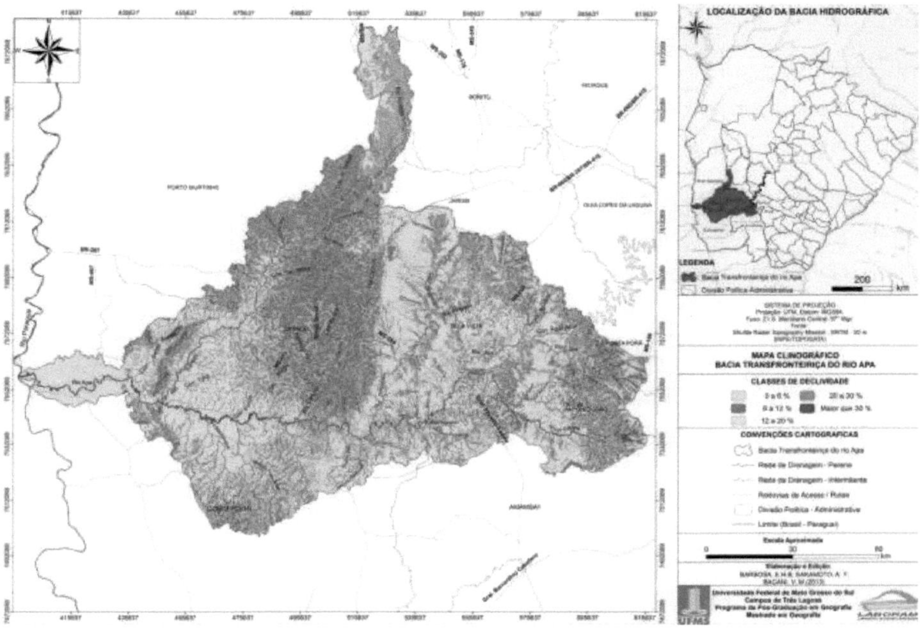

Figure 12. Clinographic map of the Apa River transboundary basin,

Table 6 quantifies the spatialization of the slope in the Apa River basin area.

Table 6. Table of slope class areas.

Slope classes	**Area (km)^{2}**	**Area (%)**
0 a 6 %	10.378,93	67,25
6 a 12 %	3.912,84	25,35
12 a 20 %	767,42	4,97
20 a 30 %	249,89	1,62
Greater than 30%	123,65	0,80
Total	15.432,72	100

It can be seen that there is a predominance of areas with a slope of 0 to 6%, representing 67.25% of the total area of the basin, followed by classes with a slope of 6 to 12%, occupying 25.35% of the basin area.

In these predominant areas, the valley is open and fairly flat. Mendonça (1999) points out that the erosion process in these areas is mainly laminar, due to the low speed of surface runoff, and can give rise to ravines. Generally, in these low areas, water flows in sheets and extensive areas of flooding occur, forming areas of floodplain.

Also according to Mendonça (1999), the action of water on the relief is associated with other factors such as geology and soil cover, which may be more important in triggering erosion processes than the hypsometry of the area.

Due to its importance, it deserves specific and in-depth studies. Among these, the natural erosion potential is a cartographic product that determines, through a weighted overlay, the areas susceptible to erosion processes, taking into account the slope of the land and the soil cover.

The slopes of the Apa River basin are oriented dominantly to the south, characterized by low solar radiation, as shown in Figure 13.

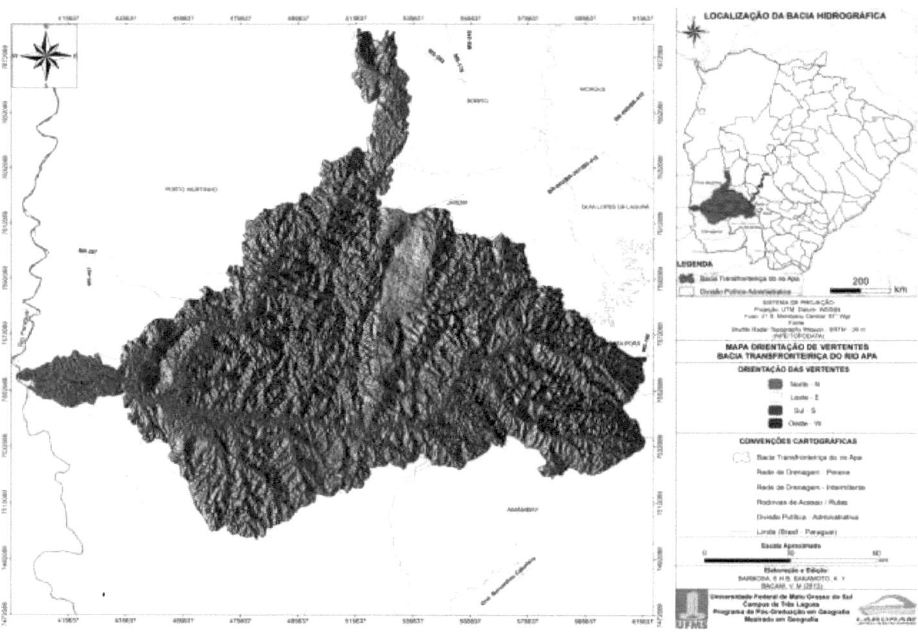

Figure 13. Slope Orientation Map of the Apa River transboundary basin.

According to Neves (1989), the west-facing slopes are the most unfavorable in terms of solar incidence and thermal comfort, as they receive direct solar radiation in the afternoon, when the temperature is highest. Thus, the right bank receives more solar radiation.

With regard to the east-facing slopes, Neves (1989) states that they are better in terms of thermal comfort, being mainly favorable for urban occupation, since the sun shines directly on them in the morning, when temperatures are milder. They are therefore suitable for projects aimed at restoring the forest, especially in areas of permanent preservation. The areas that receive the least amount of solar heat are distributed in the upper reaches and throughout the basin.

In the southern hemisphere, north-facing slopes receive more solar irradiance, while south-facing slopes receive less solar incidence and are therefore less privileged in terms of sunlight. According to Lombardo (1996), the flow of radiation arriving at a steeply sloping slope facing north in southern subtropical areas will be more intense than another sloping slope facing south in the same location.

In this sense, the slope orientation map has proved to be an important document for guiding future occupations, forest management and agricultural production in the transboundary basin of the Apa River.

6.2. Land Use and Vegetation Cover

The temporal evaluation of land use and vegetation cover revealed the various production and transformation techniques in the territory of the transboundary Apa River basin.

Seven thematic categories were classified for 1987 and eight classes for the 2012 mapping. The following classes were identified: Bodies of Water, Wetland, Forest Fragment, Exposed Soil, Pasture, Cultivated Areas (Agriculture), Silviculture (Eucalyptus) and Urban Area. The 1987 map did not take into account the Forestry (Eucalyptus) class, due to the lack of support for validating the class, and the thematic class called Forest was used to characterize all areas with dense tree cover.

The recently mapped classes (2012) allowed for field reconnaissance on the Brazilian side of the analyzed basin.

The thematic class called Bodies of Water corresponds to perennial water bodies with a width of approximately 50 to 70 meters (Figure 14) and water mirrors with a minimum surface area of 1.00 ha, called natural lakes or lagoons.

Figure 14. The Apa River, a stretch bordering Brazil and Paraguay, located at the Army detachment of the 2ª Cia. de Fronteira.

The Wet Area class corresponds to flood areas with intermittent or permanent characteristics. Figure 15 shows an area of ebb and flow.

Figure 15. Area of ebb and flow, on the stretch of MS 384.

According to FADEMS (2012), the hydrographic basin lies within the Cerrado biome with great influence from the Chaco and the Atlantic Forest.

The Forest Fragment class has the typical vegetation cover of Cerrado (open woodland), Cerradao (dense woodland), gallery forest and hillside forest, as shown in Figure 16.

Figure 16. Cerradoo, on the stretch of MS 384.

Areas of unprotected soil were observed, with no vegetation cover. In order to open the roads that give access to the interior of the Apa River basin, it was necessary to suppress the vegetation in the area, as shown in Figure 17. Soil exposure occurs in pasture areas, where the process is triggered by the acquisition of new pasture areas, deforestation, lack of soil preparation and management, and extractive activities that modify the natural landscape. The thematic class Exposed Soil has been named after this characteristic.

Figure 17. a) Earthmoving for landfill purposes, located on the stretch of the MS 166 and, b) Maintenance of the stretch of the MS 384 road.

The main production and transformation activity in the territory of the basin analyzed is livestock farming, represented by the Pasture class, where beef, dairy and extensive livestock farming activities are carried out, as shown in Figure 18. The pasture is made up of *Brachiaria decumbens* and *Brachiaria humidicula,* found in wet areas with low slopes.

Figure 18. **Figure 18. a) Beef cattle on a rural property and b) extensive cattle farming.**

The Crop Area (Agriculture) class, exemplified in Figure 19, are areas used for extensive agriculture, found mainly in the eastern region of the Apa River basin. The location of production is related to the basin's proximity to the state's grain-producing regions.

Figure 19. a) Maize plantation with soybeans in the background and b) Bean plantation on a small property.

The Silviculture (Eucalyptus) class is represented by another form of monoculture using eucalyptus, shown in Figure 20. This class is more prevalent near areas of extensive agriculture.

Figure 20. Areas with monoculture (eucalyptus), on the stretch of the MS 270.

The class called Urban Area (Figure 21) refers to the urban areas present in the Apa River basin, represented by the cities of: Antônio Joâo, Bela Vista, Caracol and in Paraguayan territory by *Bella Vista, Concepción and San Carlos.*

Figure 21. Urban area of the municipality of Caracol.

Analyzing land use and vegetation cover in 1987 and 2012, it can be seen that in 25 years the landscape of the Apa River basin has been heavily modified by anthropogenic actions. Figures 22 and 23 show the spatialization of the different types of land use and cover in the territory of the transboundary Apa River basin.

Table 7 shows that the thematic mapping of land use and vegetation cover of the Apa River carried out using the Battacharya classifier was of very good quality for 2012.

Table 7. *Kappa* Validation and Global Accuracy for 2012 for the Apa River transboundary basin.

Classifier		Global Accuracy	Quality*
Battacharya	***Kappa* index** 0.73	0,76	Very good

*According to Landis and Koch (1977).

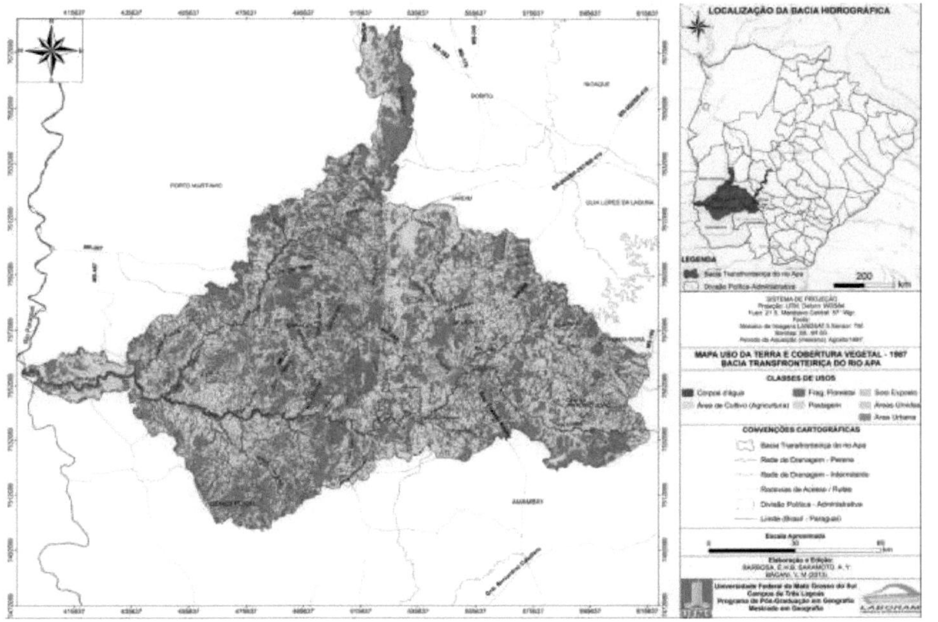

Figure 22. Land use and vegetation cover map for 1987 of the Apa River transboundary basin.

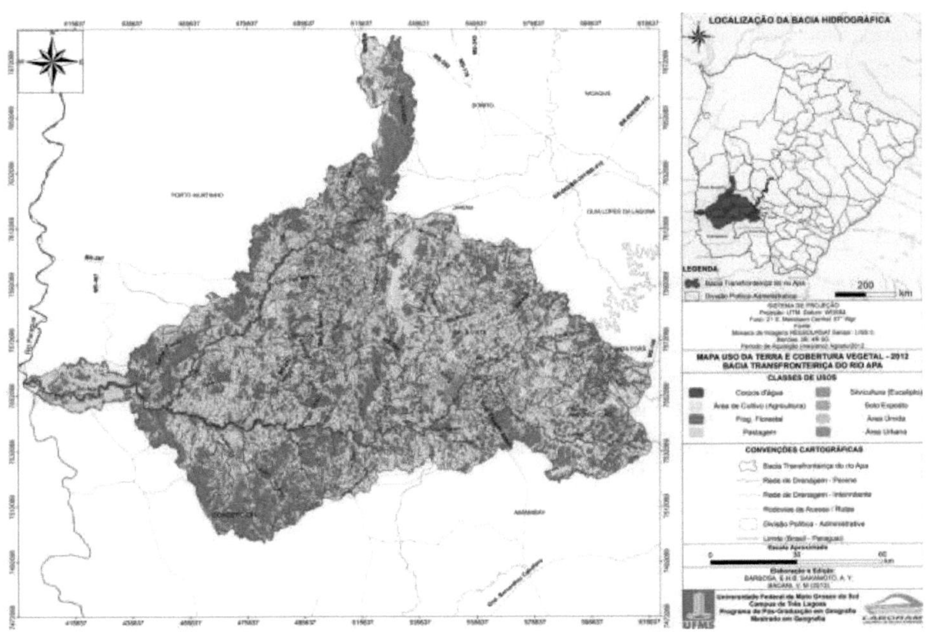

Figure 23. 2012 Land Use and Vegetation Cover Map of the Apa River transboundary basin.

Table 8 below shows the total area occupied by each mapped class over the two dates compared in the temporal mapping.

Table 8. Area occupied by thematic classes of land use and vegetation cover in 1987 and 2012.

Thematic Class	Area (km)²	Area (%)	Area (km2)	Area (%)
	1987		2012	
Bodies of water	45,27	0,29	44,44	0,29
Cultivation Area (Agriculture)	281,98	1,83	491,23	3,18
Wet Areas	906,43	5,87	686,84	4,45
Urban Area	34,71	0,22	46,85	0,27
Forest fragment	6.834,85	44,29	6.621,40	42,91
Pasture	7.162,83	46,41	7.489,73	48,53
Exposed Soil	166,24	1,08	21,16	0,14
Forestry (Eucalyptus)	*	*	36,03	0,23
Total	**15.432,30**	**100,00**	**15.432,17**	**100,00**

The classes represented by Pasture, Forest, Cultivated Area and Urban Area allowed for a comparison over the two years analyzed, as they showed significant variations over the years.

The class called Cultivated Areas represents the agricultural activities present in the basin. It can be seen that during this 25-year period the area increased by 209.25 km^2 , which corresponds to approximately 1.36 %.

Another class that represents the mode of production in the basin is Pasture. This, in turn, increased its area by 2.11 % over 25 years, corresponding to an increase of 326.90 km2.

When relating production to dense tree vegetation, which in this mapping is represented by the Forest class, it can be seen that 1.38 % of its area was suppressed, totaling 213.45 km2. As such, it is clear that the forest was suppressed in order to cultivate areas. Only 0.02% was used to increase pasture areas in the Apa River basin.

Urban areas in the Apa River basin increased by 6.62 km2, due to population growth. The municipality of Bela Vista had the largest urban expansion, with an increase of 7.74 km2 in area.

In the mapping of both years, exposed soil is evident near the drainage network. This is related to fluvial accumulation processes, which may be natural due to the shape of the

Tabular relief and accelerated by the removal of vegetation cover near the drainage network of the Apa River basin. In contrast, the Exposed Soil class showed a decrease of 0.94 %, a fact explained by agricultural practices and reforestation activities.

The reforestation activity, using Eucalyptus, is inserted in the basin, but a temporal analysis of the activity was not possible due to the years chosen.

Despite not showing significant or alarming results, the Apa River basin has undergone changes over this 25-year period in order to meet production needs. Thus, from a socio-economic point of view, this is an area with little diversification of land use, maintaining the state's economic tradition in terms of products and land structure, fitting in with the understanding that the economy of Mato Grosso do Sul is based on the cattle-soybean binomial, although the advance of forestry on the edge of the Apa and agro-energy in areas not far from this unit is noticeable FADEMS (2012).

Therefore, it is clear from this temporal mapping that the main economic activity in the Apa River basin is extensive livestock farming and extensive agriculture, in order to serve the consumer market. The tendency is to increase the need and search for new areas in order to increase the production of grain and wood, using planted forests.

The temporal evaluation identifies this change in the environmental scenario of the analyzed basin. Over the years, there has been a need to increase production in order to serve new consumer markets. As a result, techniques are perfected and technological changes are made. One example of this in the rural environment is the mechanization of the countryside.

6.3. Mapping and Applications of Environmental Legislation

Environmental legislation is a political determination with coercive purposes, aimed at preserving and conserving the natural environment in order to maintain the balance and natural maintenance of the species found there.

The transboundary basin of the Apa River has a peculiarity compared to other river basins, due to the formation of a mosaic of Conservation Units on both sides (Brazilian and Paraguayan), as shown in Figure 24.

On the Brazilian side there are 07 (seven) Conservation Units within the Apa River basin, classified according to Article 7 of Federal Law No. 9.985 of July 18, 2000, which established the National System of Conservation Units - SNUC, as Integral Protection Units and Sustainable Use Units. On the Paraguayan side are the *Paso Bravo* National Park and the *Bella Vista* National Park, the latter of which is less representative of the transboundary basin of the Apa River.

The Conservation Units were created in 2000, the first being the Private Natural Heritage Reserve - RPPN Margarida, by means of Ordinance 034/2000, published in the Official Gazette on June 16, 2000. In the same year, the Federal Decree of September 21, 2000, created the Serra da Bodoquena National Park. Municipal decrees then created the Environmental Protection Areas (APA) and the only Municipal Natural Park.

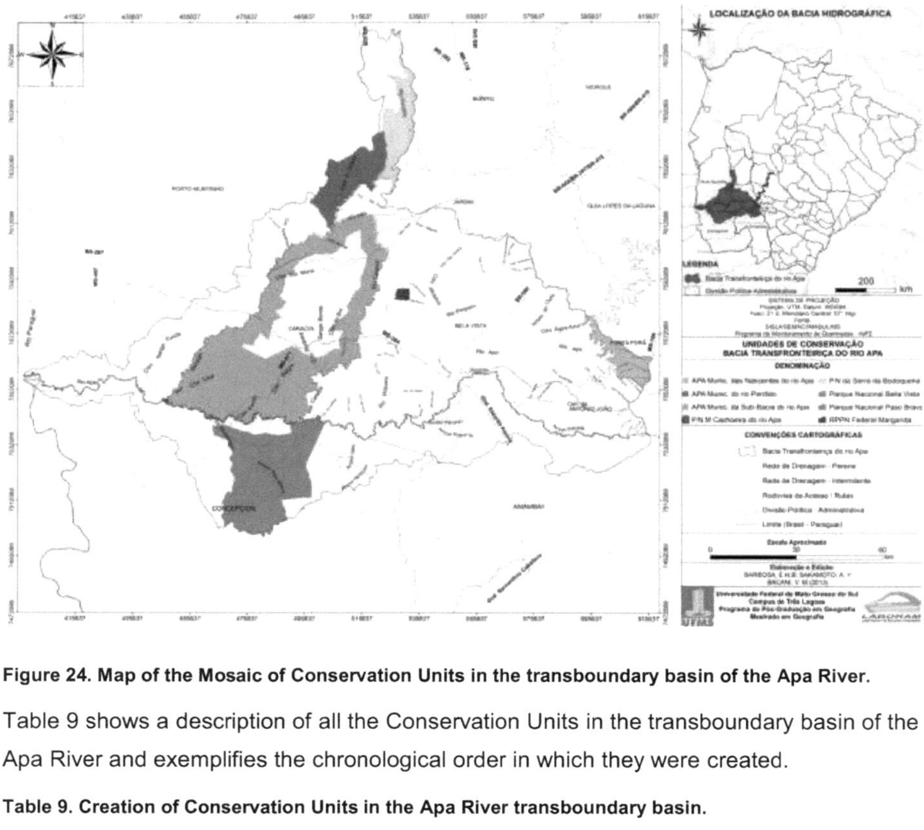

Figure 24. Map of the Mosaic of Conservation Units in the transboundary basin of the Apa River.

Table 9 shows a description of all the Conservation Units in the transboundary basin of the Apa River and exemplifies the chronological order in which they were created.

Table 9. Creation of Conservation Units in the Apa River transboundary basin.

Conservation Unit	Act of creation	Total area (ha)	Area within the basin (ha)
Margarida Federal RPPN	Ordinance no. 034/2000	1.999,31	1.999,31
Serra da Bodoquena	Federal Decree of September 2000	76.826,13	28.666,46
Municipal Natural Park of	Municipal Decree no.	58,31	58,31

Apa River Waterfall APA Municipal	2.573/2001		
Springs	Municipal Decree no.	17.196,16	16.154,90
do Rio Apa APA Municipal do Rio	4.743/2004		
Perdido	Municipal Decree no.	36.246,02	36.246,02
	3.107/2005		
Sub-Basin Municipal APA	Municipal Decree no.	195.148,52	195.148,52
of the Rio Apa *Paso Bravo*	052/2009		
National Park	*Decree of Power*	103.108,00	97.380,95
	Executive Order No. 20.712 of April 20, 1998		
***Bella Vista* National Park**	-	7.261,70	19,59

It can be seen that only 37.31% of the Serra da Bodoquena National Park falls within the basin area. The APA Municipal das Nascentes do Rio Apa has 93.94% of its area within the basin. The other PAs are 100% within the basin.

In the transboundary basin of the Apa River, the largest area occupied by a Conservation Unit is the APA of the Apa River Sub-Basin, with an area of 195,148.52 ha.

The preservation areas in the Apa River basin, for legislative purposes, are called Permanent Preservation Areas (APPs). PPAs are protected areas, whether or not they are covered by native vegetation, with the environmental function of preserving water resources, the landscape, geological stability and biodiversity, facilitating the gene flow of fauna and flora, protecting the soil and ensuring the well-being of human populations, according to item II of Article 3 of Law No. 12,651 of May 25, 2012.

For the transboundary basin of the Apa River, in accordance with the New Brazilian Forestry Code, the following areas apply to any natural perennial and intermittent watercourse, excluding ephemeral watercourses, from the edge of the channel of the regular bed:

- 30 (thirty) meters, for watercourses less than 10 (ten) meters wide;
- 50 (fifty) meters, for watercourses between 10 (ten) and 50 (fifty) meters wide;
- 100 (one hundred) meters, for watercourses between 50 (fifty) and 200 (two hundred) meters wide.

Areas around natural lakes and lagoons, with a minimum width of:

- 100 (one hundred) meters, in rural areas, except for bodies of water of up to 20 (twenty) hectares, whose marginal strip will be 50 (fifty) meters;

- 30 (thirty) meters in urban areas.

The other areas are:

- Areas around springs and perennial waterholes, regardless of their topographic location, within a minimum radius of 50 (fifty) meters;

- On hilltops, hills, mountains and mountain ranges, with a minimum height of 100 (one hundred) meters and an average slope greater than 25°, the areas delimited from the contour line corresponding to 2/3 (two thirds) of the minimum height of the elevation always in relation to the base, the latter being defined by the horizontal plane determined by the adjacent plain or body of water or, in undulating terrain, by the elevation of the saddle point closest to the elevation.

In Paraguay, *Law No. 422/73 Forestry,* in its Article 2, point "d", aims to protect water basins and springs.

Paraguayan law does not determine preservation strips along watercourses, hilltops or slopes above 45°. The only similarity is the determination of areas within rural properties, which are similar to the legal reserve areas determined by Brazilian law.

Thus, *Forestry Law No. 422/73* states:

Art. 42: *All rural properties of more than twenty hectares in forest areas must maintain twenty-five percent of their area in natural woodland. In the event of not having this minimum percentage, the landowner must reforest an area equivalent to five percent of the property's area.*

Therefore, in order to determine the mapping of environmental legislation in the transboundary basin of the Apa River, the spatialization of land in the two countries must be taken into account.

Source mapping would be one of the tools for determining the forest recovery of permanent preservation areas with consolidated areas, since the New Brazilian Forest Code takes into account the property's fiscal module for the recovery of APPs, according to Art. 61.

Figure 25 exemplifies the spatialization of Permanent Preservation Areas, determined by Brazilian Federal Law No. 12.651 of May 25, 2012.

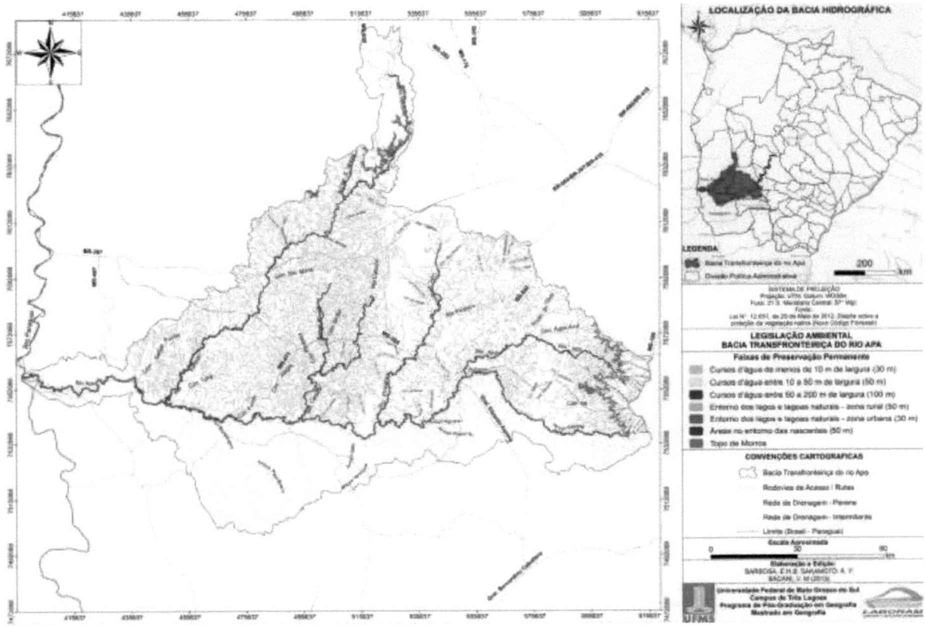

Figure 25. Map of Permanent Preservation Areas on the Brazilian side of the Apa River transboundary basin.

Table 10 shows the areas set aside for permanent preservation on the Brazilian side of the Apa River basin.

Table 10. Permanent Preservation Areas in the Apa River basin. Description Area to be determined (km)²

Description	Area to be determined (km²)
30 meters, for watercourses less than 10 meters long width	417,28
50 meters, for watercourses between 10 and 50 meters long width	23,06
100 meters, for watercourses between 50 and 200 meters long width	259,33
Around lakes and natural ponds - 50 meters in rural areas	6,80
Around lakes and natural ponds - 30 meters in urban areas	0,07
Areas around springs and perennial waterholes, whatever its topographical location, within a radius of at least	19,29

50 meters	
On the tops of hills, mountains and mountain ranges	109,82
Total	**835,63**

An analysis of the APPs shows that 5.14 % of the total area of the Apa River basin should be set aside for permanent preservation. Of this total, perennial and intermittent streams generate 2.70% of APPs of 30 meters, while stretches of the Caracol River, the Boi stream, the Piripucu River, the Agua Azul stream, the Santa Maria stream, the Taquaruçu stream and the Ita stream have APPs that occupy 0.15% of the basin's total.

The Apa, Perdido, Caracol, Piripucu and Estrela rivers make up an area of 259.33 km^2 of APP, which represents 1.68 % of the total area of the basin.

The hilltops are located on the escarpments of the Serra de Xaraxaxà, Serra da Bodoquena and Morro do Teima, occupying an area of 109.82 km2.

The areas around lakes and natural ponds combined (urban and rural) occupy 6.87 km2. In the urban area, lakes were found in the municipality of Bela Vista.

The other areas are generated by the springs present in the water system on the Brazilian side, with an area of 19.29 km2, which represents 0.12 % of the total area of the basin.

Of these APPs, 508.43 km2 are covered by dense tree vegetation, characteristic of riparian or gallery forest areas. This means that 60.84 % of the APPs are preserved. Figure 26 shows the vegetation cover present in the APPs.

The remaining 35.36% are areas of conflict[2] with the type of use and occupation of rural properties in the basin. Table 11 shows the cross-referencing of APPs with the type of land use and vegetation cover.

Table 11. Types of Land Use in conflict with Permanent Preservation Areas in the Apa River basin.

Land Use Classes (2012)	**Area affected by APPs (km^2)**
Wet Areas	89,01
Pasture	186,60
Crop Areas	14,20
Exposed Soil	2,46

2 The difference of 3.80% was excluded from the analysis as it represents the thematic class "water" and "forest", made up of rivers and streams with a width of approximately 50 to 70 meters.

Reforestation	0,31
Urban Area	2,91
Total	**295,51**

The class identified as "Wet Areas" is generally used as pasture during the dry season. The mapping prioritized the spectral response of the area, but in reality, these mapped classes are used for livestock.

Through this cross-tabulation of data, it is clear that the two classes that represent production and occupation in the basin affect the APPs. The class Pasture is the most representative, with 22.33%, and the class Crop Area represents 1.70% of occupation in the APPs. Use and occupation encroaches on 24.03% of the total allocated to APPs.

Exposed Soil is related to the presence of sand banks or the formation of islands along the watercourse, especially in the Apa River, where this class is predominant.

The Urban Area had an incidence of 0.35 % in the APPs, which is due to urban expansion in the municipalities of Bela Vista, Antonio João and Caracol.

Figure 27 shows the incidence of the type of land use in the APPs on the Brazilian side of the Apa River basin.

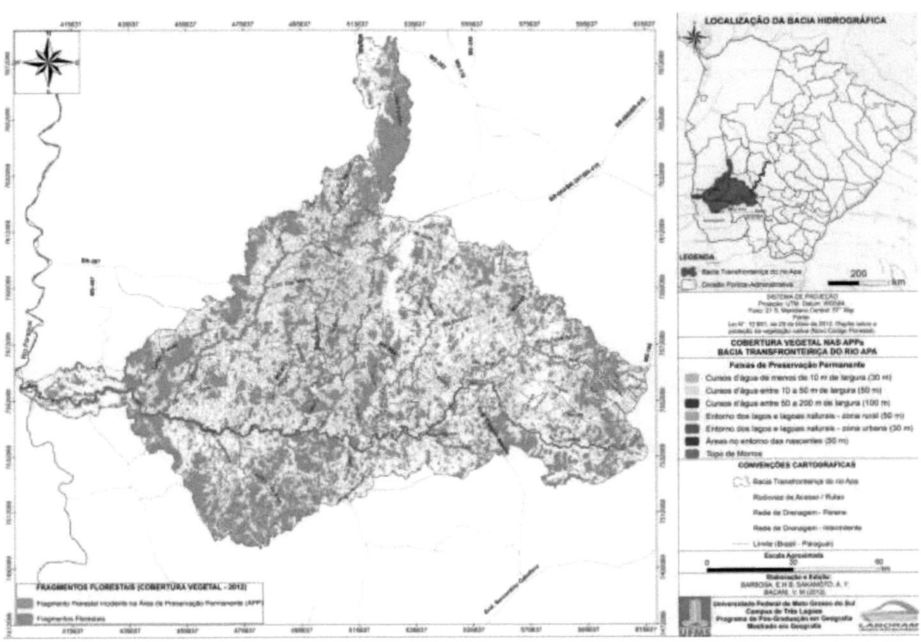

Figure 26. Map of the vegetation cover present in the APPs.

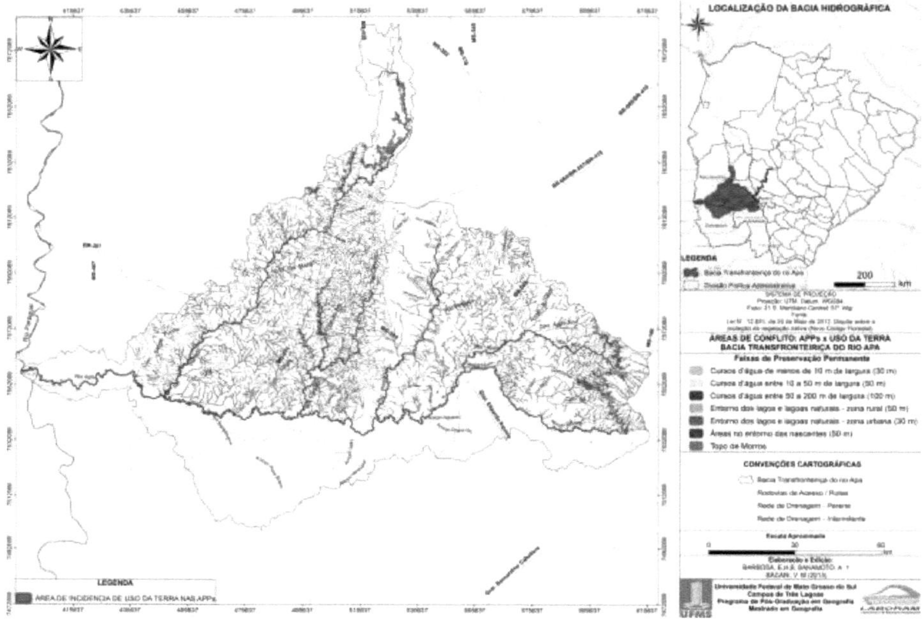

Figure 27. Map of Land Use in conflict with Permanent Preservation Areas in the Apa River basin

6.4. Environmental Fragility Analysis

6.4.1. Natural Erosion Potential

The natural erosion potential is a product that anticipates the analysis of environmental fragility. This cartographic product indicates areas where erosion processes are natural, due to the type of soil, rainfall intensity, topographical characteristics and the biological importance of the environment.

The Apa River basin is structured on a predominantly flat terrain with a range of slopes from 0 to 6% and the presence of hydromorphic soils with a texture varying from sandy to clayey. Due to these characteristics, water flows in sheets and extensive areas are flooded, forming areas of floodplain. The erosion process is mainly laminar, due to the low speed of surface runoff, and can give rise to ravines.

Areas of flat terrain (range 0-6%) and sandy and/or clayey soils have a low natural potential for erosion. This represents 39.74 % of the total area of the Apa River basin, so under conditions of inadequate management the erosion process (gullies) can be enhanced.

However, the map of natural erosion potential or emerging fragility is a summary of the natural erosion phenomena within the transboundary basin of the Apa River, as shown in Figure 28.

The distribution by area of the Natural Potential for Soil Erosion classes is described in Table 12.

Table 12. Distribution of Natural Erosion Potential classes.

Classes	**Area (km)^{2}**	**^Area^%^**
Low	6.132,29	39,74
Average	7.254,12	47,00
High	2.046,32	13,26
Total	**15.432,72**	**100,00**

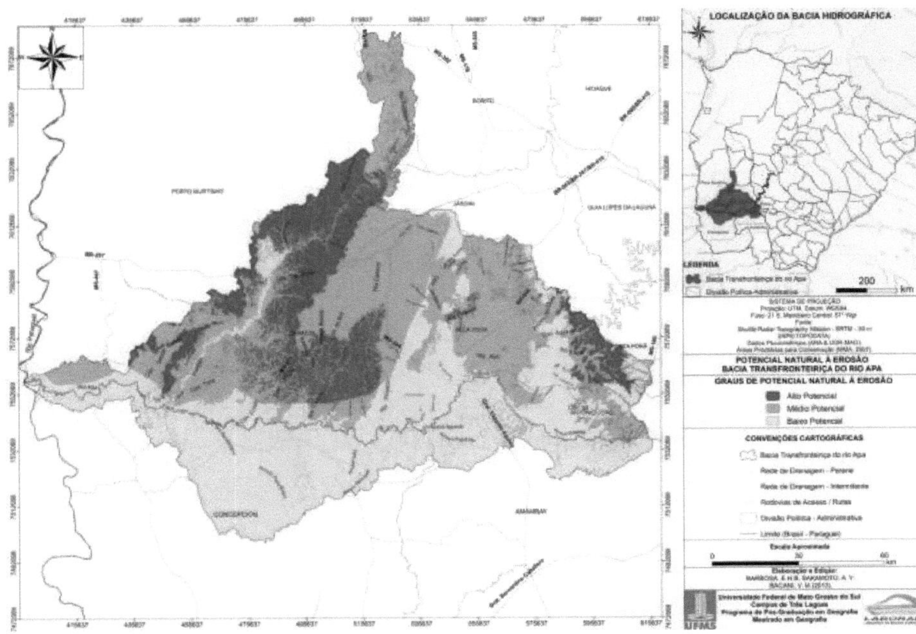

Figure 28. Map of Natural Erosion Potential of the Apa River transboundary basin,

The Low class is found in 39.74 % of the study area, associated with gentle relief with flattened forms that do not exceed a slope of 20%, with altimetric variation from 90 to 690 m, where clay-textured Latosols, clay to sandy-textured Red-Yellow Argisols and clay-textured Vertisols develop. Rainfall is concentrated in these areas and ranges from

1,376.8 to 1,578.7 mm per year. Data on biological importance was not attributed, as there was insufficient information known, which reduced the degree of fragility.

It can be seen that 47.00 % of the basin analyzed is considered to be of medium potential fragility to erosion, located in the central region of the basin, bordered to the north by the Serra da Bodoquena, to the south by the Apa River, to the east by the Caracol River and to the west by the basin boundary near the Perdido River.

In the area considered Medium, the predominant soils are: Argissolo Vermelho-Amarelo with a clayey to gravelly texture, Planossolo Nâtrico with a sandy/medium and medium texture, Neossolo Regolitico with a medium very gravelly texture and Chernossolo Rêndzico. These are characteristic soils with slopes ranging from 6% to 30%. Rainfall in this area ranges from 1,246.5 to 1,578.7 mm per year. The biological importance of this area varies from High to Extremely High, due to the presence of the Serra da Bodoquena National Park.

High fragility corresponds to 2,046.32 km^2 , the smallest representation in the basin, with approximately 13.26 %, with slopes greater than 30% associated with lithic and regolithic neosols with a medium sandy/gravelly to very gravelly texture, a region of extreme biological importance.

Thus, the Apa River basin has 60.26 % of natural erosion potential, which varies from high to very high, precisely in areas sustained by dissected relief and soils with sandy and gravelly characteristics.

6.4.2. Environmental Fragility

The mapping of environmental fragility indicated the areas susceptible to imbalances caused by the process of land use and occupation. The mapping allows a perception of the landscape as a whole, indicating the degrees of fragility related to soil protection.

After observing the characteristics of the terrain in the face of erosive processes caused by rainfall, the pedological characteristics and biological importance of the sectors susceptible to erosion in the Apa River basin, the analysis seeks to synthesize the erosive processes related to the use and production carried out in the territory of the Apa River basin.

Combining the map of natural potential for soil erosion with the map of land use and vegetation cover analysis from 2012 resulted in a synthesis map called environmental fragility, as shown in Figure 29.

Table 13 shows the quantification of the distribution of environmental fragility classes found in the transboundary basin of the Apa River.

Table 13. Distribution of areas of Natural Fragility.

Classes	/,. Area (km)^2	Area (%)
Low	2.584,51	16,75
Average	7.293,46	47,26
High	5.469,74	35,44
Very high	85,01	0,55
Total	**15.432,72**	**100,00**

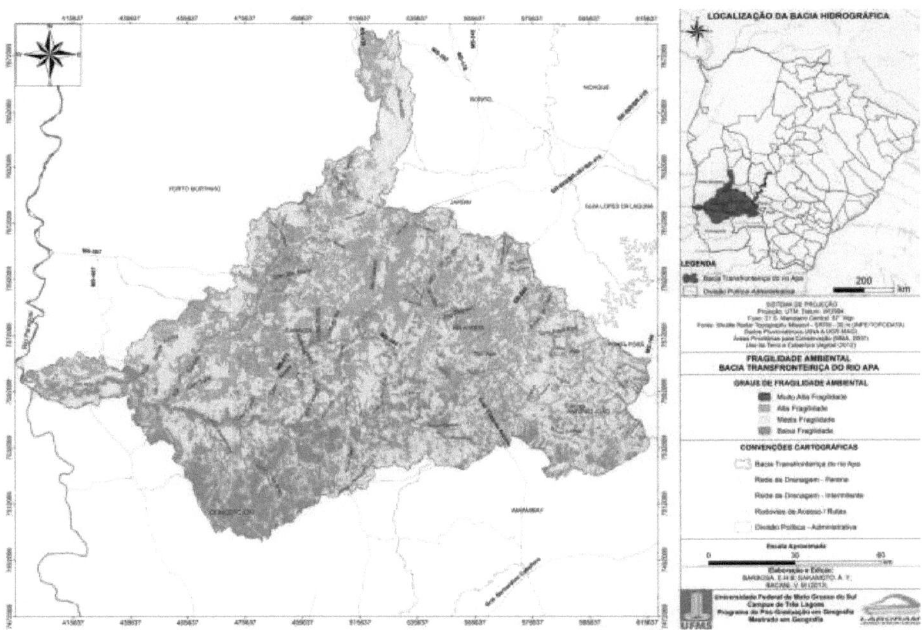

Figure 29. Environmental Fragility Map of the Apa River transboundary basin

Low environmental fragility is found in the 145.64 km stretch from the middle course of the Perdido River to the mouth of the Apa River, from the banks of the Apa River to the confluence with the Pirarucu River, and on the Paraguayan side of the basin occupies 46.19 % of the total area. This result is related to the low potential fragility and the presence of natural vegetation that has the role of protecting the soil, located in the riparian forests.

Medium environmental fragility is predominant in the transboundary river basin, occupying an area of 7,293.46 km^2 , which represents 47.26 % of the total basin area. This class is spatialized in areas destined for pasture and in areas defined as wetlands.

High environmental fragility has the second highest spatial distribution with 5,469.74 km2 (35.44 % of the basin area), and is directly related to anthropogenic action caused by the clearing of natural vegetation to introduce pasture for farming and agriculture, since a large part of this area is in a region classified as extremely high by the biological importance map.

When it comes to use, occupation and production in the basin, it can be seen that the Conservation Units are predominantly highly fragile, due to the lack of proper management of pasture areas, causing erosion processes and contributing to the silting up of the basin's water sources. In these areas where sandy soils predominate, conservation techniques should be adopted to minimize water erosion problems. This integration results in the addition of organic matter to the soil, improved structure, increased water infiltration, an increase in the soil's production capacity and a reduction in the speed of runoff.

The areas considered Very High correspond to an area of 85.01 km2. It is characterized by the relationship between the classes water, wetlands, urban area and exposed soil with very high potential fragility to erosion, since these factors are associated with high slopes, corresponding to over 30%, susceptible to erosive processes, compromising water and edaphic resources.

6.5. Environmental Zoning Proposal

After determining the areas of environmental fragility, it is necessary to implement environmental conditions for conservation purposes in order to minimize the negative impacts caused by human activity.

The economic activities developed in the transboundary basin of the Apa River are extensive livestock farming and agriculture. There are a number of agrarian reform settlements (smallholdings) which set aside small areas for planting crops.

The main problems detected in the hydrographic basin analyzed come from the mode of production, among them: erosion, pollution and water contamination.

Erosion is associated with a lack of proper management in areas where farming takes place and in some stretches on the banks of roads that give access to the interior of the basin.

The basin has four urbanized areas, so water pollution and contamination are caused by the lack of basic sanitation and disorderly occupation of the banks of watercourses.

Therefore, the Environmental Prognosis is presented, which defines the environmental zones and sets out the recommendations for the Environmental Zoning Proposal for the transboundary basin of the Apa River, described in Table 3.

This study is based on the legal precepts (Brazilian Forest Code and *Forest Law No. 422/73*), the spatialization of environmental fragility classes and information on land use and vegetation cover in 2012, which made it possible to establish different degrees of restriction on land use.

Table 3. Environmental forecast.

Zone	Category	Criteria	Recommendations
Legal Restriction Zone - LRZ	APPs Municipal Nature Park National Park	Environmental legislation Integral Protection Conservation Units	- Isolation of the APP area, preventing access by animals and other production and/or occupation activities; • Planting typical native cerrado tree species around water courses; • Maintain forest remnants, except in cases provided for by specific law.
	Protected wilderness areas	*Forestry Law no. 422/73*	- *Comply with Art. 42 of La Forestry Law no. 422/73.*
	Forest Remnants	Forest Fragments	-Plant suppression is permitted in the cases provided for by specific law and by registering the respective legal reserve on the property.
Restricted Occupation Zone - ZOR	Hilltops	Environmental legislation.	- Isolation of the area, preventing access by animals and other production and/or occupation activities.
Urbanized Zone - ZU	Urbanized Area	Homogeneous built-up area.	• Determine the floodplain, prohibiting occupation in this area and develop environmental education projects with residents. • Draw up Master Plans and/or

			revise existing ones (municipality of Bela Vista - MS).
Zone of Incongruence - ZI	Conflict between land use and environmental legislation	Combining APPs with anthropized areas	- Recovery of riparian forest, in compliance with Law No. 12.651 of May 25, 2012.
Unit Zone Conservation Area - ZUC	Environmental Protection Area - APA	Municipal Decree	Comply with respective Management Plans (approved by the government).
	Private Natural Heritage Reserve - RPPN	Ordinance	
Productive Zone Rural - ZPR	High Restriction	- Slope between 20 and 30%. • Red-Yellow Argisol. • Pasture areas.	• Curve determination leveling and terracing. • Indicated use: livestock with limitations and reforestation.
	Average Restriction	- Slope between 6 and 20 %. • Red-Yellow Argisol. • Areas of paste ɑ e		• Encouraging agriculture family. Agricultural use with intensive breeding. • Keep the fragments of forests.

Continued...

Zone	Category	Criteria	Recommendations
Area of Settlement Rural - ZAR	Agrarian Reform Settlement.	Family Farming	• Encouraging family farming; • Public authorities offer Rural Extension, through the competent bodies; • Agricultural use with intensive breeding; • Maintain forest fragments. • Recover the riparian forest, observing the provisions of Law No. 12.651 of May 25, 2012.
Indigenous Lands	Indigenous Lands	SPU registration.	• Encouraging family farming;

Zone - ZTI	Approval decree published.	• Public authorities offer Rural Extension, through the competent bodies; • Agricultural use with intensive breeding; • Maintain forest fragments. • Recover the riparian forest, observing the provisions of Law No. 12,651 of May 25, 2012.

In addition to the guidelines presented in the Environmental Forecast, the priorities determined in the Priority Areas for the Conservation, Sustainable Use and Benefit Sharing of Brazilian Biodiversity (Ordinance No. 9 of January 23, 2007 in view of the provisions of Decrees No. 2,519 of March 16, 1998 and No. 5,092 of May 21, 2004) will be taken into account. This defines the following areas:

a) Creation of the Serra do Maracaju Corridor: recovery of the springs and formulation of the Creation of a Conservation Unit for Integral Protection and Sustainable Use, since there is a strong process of silting up of the rivers; charcoal factories; farming; sugar cane plantations; inadequate management of APPs;

b) Bodoquena Plateau: Encouraging the creation of RPPNs, as the area divides the waters of the Serra da Bodoquena; recharge area for local aquifers; border area with the Pantanal region.

c) Foz do Apa: Recovering the riparian forests of the Apa River. Due to the presence of patches of dry Chaco (hawthorn), a region where Carandâs and Paratudo are concentrated, and because it is a landscape complex influenced by Chaco, Cerrado and semi-deciduous (seasonal) dry forests, it is necessary to create a Natural Park type CU.

The recovery of the riparian forest in the APPs and Hilltops should be subjected to the Forest Enrichment process, according to the methodology proposed by Durigan et. al. (2011).

The choice of species should be based on the work carried out by Pott and Pott (2003). The choice will prioritize native species that are preferably present in locations with the same characteristics as the area to be reclaimed.

The summary of the environmental prognosis is the environmental zoning, Figure 30 corresponds to the map of the environmental zoning proposal for the transboundary basin of the Apa River.

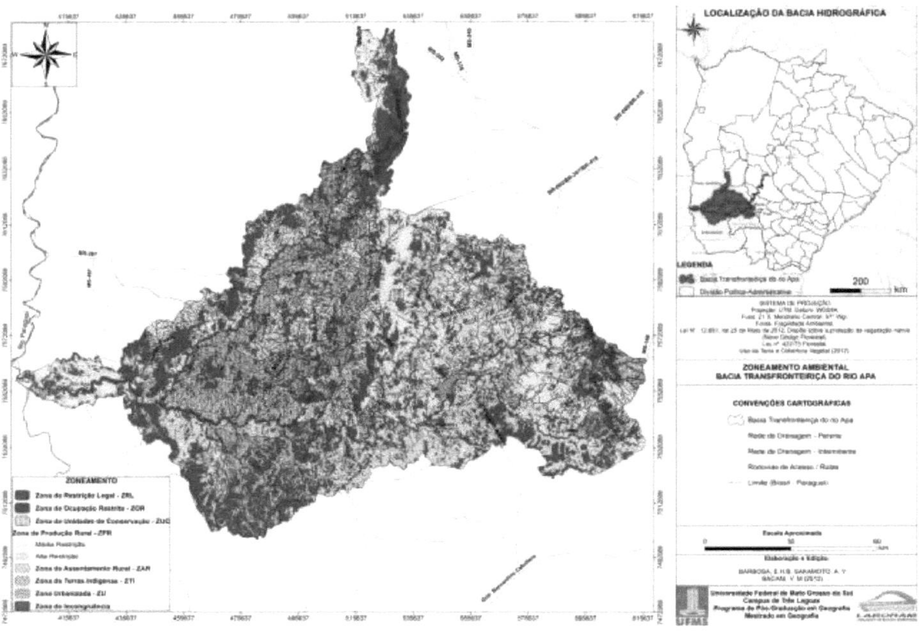

Figure 30. Environmental zoning map of the Apa River transboundary basin.

The areas indicated for recuperation, restriction of occupation and preservation should be isolated with the installation of smooth wire fencing, with treated eucalyptus or concrete mortar. The purpose of the fence is to prevent animals from accessing the APPs.

The preparation of the areas must respect the fragility and potential of the landscape, without interfering with the geodynamics of the site to be reclaimed. Most of the areas can be modified by anthropogenic means, such as pastures, agriculture and areas of exposed or compacted soil. Soil preparation practices may vary according to the reality of each area to be reclaimed.

The aim is to make the soil suitable for the growth of seedlings by eliminating all cultural waste through root digging, among other practices that promote soil cleaning.

Soil preparation practices are described below:

- Selective manual mowing to control weeds or invasive plants, especially grasses present on the land because they are aggressive and also serve as a focus for the spread of pests and diseases;

- Placing the planting lines on a level, using devices suitable for leveling, according to the established spacing;

- Mark the holes with a hoe, slats, ropes or pickets;
- Coveamento;
- Crowning;
- Organic fertilization;
- Preservation of the fog;
- Weeding between planting lines;
- Mulching.

Native plant species that are in a state of regeneration should be maintained at the planting sites, as well as the fruit trees found by trying to make holes in the remaining spaces.

The type of soil found, whether dry, waterlogged or stony, must be observed to determine the best species to plant. The soil is the main factor that determines the species that will occupy the site, considering fertility, depth and humidity.

The spacing for planting the seedlings should be 3x3 meters, three meters between rows and three meters between plants, so that the plants grow individually. The rows should follow the contour lines to minimize the risk of water erosion.

Planting must be carried out in the rainy season (November to April) on rainy days or after rains, in sufficient volume to provide soil moisture levels favorable to the adaptation of the seedlings in the field.

To prevent fire from entering the areas to be reclaimed, firebreaks should be built by weeding both sides of the fences, with a minimum width of 5 meters. This protection measure should be adopted in APPs and other areas subject to preservation or environmental regeneration.

The vegetation suppression of the areas indicated as "Forest Fragments" must be registered as a Legal Reserve on the rural property and in accordance with SEMAC/MS State Resolution No. 18 of August 5, 2008.

When analyzing urban issues, only the municipality of Bela Vista has a Master Plan (Complementary Municipal Law No. 019/2006, of October 10, 2006). Bela Vista's plan has not yet been published for revision. The other municipalities will have to draw up their respective plans in order to dimension the urban expansion of their respective municipalities.

In this way, the Environmental Prognosis enabled the Environmental Zoning for the transboundary basin of the Apa River to be drawn up, listing the use and occupation and legislative aspects. It should be noted that the management of the basin will have to be shared between the two countries, guaranteeing the efficiency of the environmental planning and management system.

This work could help formalize a mechanism for cooperation, articulation and coordination between Brazil and Paraguay for the integrated management of the Apa River basin, under the aegis of national and bilateral institutional competencies, treaties and agreements existing in the La Plata Basin. In addition to generating a Work Plan for the promotion of integrated and articulated management of the Apa River basin, through coordinated action by the bodies responsible for environmental and water resource management in Brazil and Paraguay, with the active participation of regional and local bodies and civil society, comprising short, medium and long-term actions.

7. FINAL CONSIDERATIONS

The survey of physical aspects combined with remote sensing data and geoprocessing techniques helped to obtain and tabulate all the physical elements analyzed, and were able to generate qualitative and quantitative results expressed in thematic maps, tables and charts. This generated a database of primary and secondary data, which facilitated decision-making by analyzing various data sources.

The planialtimetric base obtained from the SRTM 30 m RADAR image made it possible to classify the relief aspects of the Apa River basin, and was also important in comparing the use and occupation of the basin area. In addition, the use of this orbital product made it possible to delimit the Apa River basin correctly, on an appropriate scale for analysis and division according to altimetric classes.

The creation of thematic maps made it possible to obtain results that will contribute to the establishment of an environmental diagnosis and a physical-conservation prognosis. These thematic products, together with spatial data, fieldwork and theoretical references, including the geosystem, made it possible to determine the present physical and conservation diagnosis, which culminated in the Environmental Zoning of the Apa River basin.

The Apa River basin has been subject to anthropogenic intervention since before the Paraguayan War. As the exploration process requires new areas of intervention, the landscape is strongly altered by the production of space.

The temporal assessment of land use and vegetation cover revealed an accelerated process of natural vegetation being replaced by pasture areas. In addition, there has been an expansion of cultivated areas and a tendency for them to increase in the central to eastern direction of the basin area. This is related to the proximity of the basin area to the major grain producers in the state of Mato Grosso do Sul.

In order to curb illegal occupation, legal issues must be implemented on both sides of the river basin. As a result, public authorities must prioritize environmental issues and, above all, comply with legal precepts, carrying out coercive actions to guarantee the balance and maintenance of the natural environment.

The suggestion for determining public conservation policies begins with the formulation and consolidation of a transboundary river basin committee or mixed bilateral committees. There has been some progress in this direction: Decision No. 14, published in the Official Gazette on March 10, 2003, approves the Ministry of Foreign Affairs to begin discussions with the Paraguayan government to develop a work agenda to promote shared management of the Apa River Basin. The proposal should therefore be put into effect and work begun on shared management.

When it comes to transboundary resources, it is clear that the laws or regulations are different. In the case of the transboundary Apa River basin, the disparity between the Brazilian Forest Code (Law No. 12,651 of May 25, 2012) and the Paraguayan Forest Code (*Law No. 422/73) is* evident. It is proposed that the unification or reform of specific legislation for the Apa River basin would be the start of shared management, since any planning and management action must have a legislative framework.

If this action between the two governments comes to fruition, it will mark a historic milestone in the planning and management of transboundary river basins.

Another factor that demonstrates shared management in the Apa River basin is the formation of Conservation Units, which in the future will make it possible to form an ecological corridor integrating conservation in both countries.

The methodology used for this physical-conservation diagnosis, based on the combination of elements of the physical environment such as soil, relief, land use and vegetation cover, environmental legislation, erosion potential, using geoprocessing techniques, resulted in a model of extreme importance for the physical-territorial planning of the Apa River basin.

Therefore, this work is considered to be environmental zoning for the transboundary basin of the Apa River and a methodological proposal for investigating transboundary river basins, obeying geosystemic principles, rational land occupation, recovery from environmental degradation and minimizing problems of water conflict.

8. REFERENCES

ARCGIS/ARCINFO - Esri Inc. ***ArcGis Version* 10**. *Environmental Systems Research Institute Inc. New York.* 2012. 1 Cd Rom.

AB'SABER, A. **The nature domains of Brazil: landscape potential.** 3ª edition. Sâo Paulo: Ateliêr Editorial, 2003.

ALMEIDA, A. Q de; SOUZA, C. A. M de; TULLI, L. M. A. **Use of remote sensing techniques in the multitemporal analysis of deforestation in the Córrego do Galo watershed, Domingos Martins, ES**. Electronic Scientific Journal of Forest Engineering. Year VIII. No. 14. August 2009. Available at: <http://www.faef.br>

BACANI, V. M. **Geotechnologies applied to the physical-territorial planning of the upper Coxim river basin, MS.** 2010. 222 f. Thesis (Doctorate) - Faculty of Philosophy, Letters and Human Sciences, University of Sao Paulo, Sao Paulo, 2010.

BARBOSA, E. F da F. de M.; GUIMARAES, V. **Relationships between physical environmental potential and use restrictions in the Pontinha do Coxo stream basin - Camapuâ-MS.** In: MORETTI, E. C; CALIXTO, M. J. M. S. (Org.). Geography and regional production: society and environment. Campo Grande, MS. Ed. UFMS, 2003.

BELTRAME, A. da V. **Diagnosis of the physical environment of river basins: Model and application**. Florianópolis: UFSC, 1994.

BRAZIL. Ministry of Mines and Energy. General Secretariat. **RADAMBRASIL** Project. Sheet SF. 21-Campo Grande; Geology, Geomorphology, Pedology, Vegetation and Potential Land Use. Rio de Janeiro, 1982. 416 p.

BRAZIL. **Priority Areas for the Conservation, Sustainable Use and Benefit Sharing of Brazilian Biodiversity**: Update - MMA Ordinance No. 9, of January 23, 2007 / Ministry of the Environment, Secretariat of Biodiversity and Forests. - Brasilia: MMA, 2007.

BRAZIL. Law No. 12.651, of May 25, 2012. Provides for the protection of native vegetation. **Official Gazette of the Union**, Brasilia, DF, May 28, 2012.

BRAZIL. **Conservation plan for the Upper Paraguay river basin: integrated analysis and forecast for the Upper Paraguay river basin**. Brasilia: PNMA, 1997. 369p.

BROCH, S. A. O. Transboundary Water Management: The Case of the Apa Basin. Brasilia, 2008. 247 p. Thesis (Doctorate).

CÂMARA, G; MONTEIRO. V. M. A. **Basic Concepts in Geoinformation Science**. 1 ed.

Sao José dos Campos. INPE. 2001.

CIDEMA. 2003. "Assessment of the Water Resources of the transboundary basin of the Apa River - Mato Grosso do Sul". Final report. CIDEMA/SRH/MMA. Campo Grande-MS.
CHRISTOFOLETTI, A. **Modelagem de Sistemas Ambientais**. 1 ed. Edgard Blücher. Sao Paulo, 1999.

COELHO NETO, A. L. Slope hydrology at the interface with Geomorphology. In: GUERRA, A. J. T; CUNHA, S. B. **Geomorfologia: uma actualizaçao de bases e conceitos**. Editora Bertrand Brasil. Rio de Janeiro, 1994, 458 p.

CORRÊA, A. C. de B. **Evolutive Patterns of the Geosystems Integrating the Summit Surfaces of the Borborema Plateau, Northeast Brazil.** In: MENDONÇA, F. de A; LOWEN-SHR, C. L; SILVA, M da (Org.). Espaço e Tempo: complexidade e desafios do pensar e do geogràfico. Curitiba: Associaçao de Defesa do Meio Ambiente e Desenvolvimento de Antonia (ADEMADAN), 2009.

CREPANI, E.; MEDEIROS, J. S. de; HERNANDEZ FILHO, P.; FLORENZANO, T. G.; DUARTE, V.; BARBOSA, C. C. F. **Remote sensing and geoprocessing applied to Ecological-Economic Zoning and land use planning** (INPE-8454-RPQ/722). National Institute for Space Research, Sao José dos Campos. 2001.

BERTRAND, G. **Paisagem e Geografia fisica global: esboço metodològico.** Caderno de Ciências da Terra. Sao Paulo. v.13. p. 1-27, 1972.

DeBIASE, M. **The Clinographic Chart. Methods of representation and preparation.** Journal of the Department of Geography. Sao Paulo: FFLCH/USP. n.6, 1992.p.45-61.

DEAMO, J. C. M.; VALLE JUNIOR, R. F. do; OLIVEIRA, S. B. de; CAMILO, M. J.

Physico-Conservationist Diagnosis of the Boa Vista Stream Watershed, Uberlândia-Mg. Uberaba, 2008. Available

at:http://www.cefetuberaba.edu.br/paginas_html/revista/pdf/Resumo_12.pdf.

Accessed on: 19/10/2009.

DIRETORIA DO SERVIÇO GEOGRAFICO DO EXÉRCITO (DSG) - Sheet SF 21-z- A-III. **Antonio Joào.** Scale 1:100.000. Third printing, 1999.

DURIGAN, Giselda et al. **Manual para recuperação da vegetaçâo de cerrado**. 3. ed. rev. and current. Sao Paulo: SMA, 2011.

BRAZILIAN AGRICULTURAL RESEARCH COMPANY - EMBRAPA. **Brazilian soil**

classification system. National Soil Research Center. 3 ed. Rio de Janeiro, 2011.

ESSELIN, P. M; OLIVEIRA, T. C. M. indio, Gado e Armagem na Construçao da Fronteira no Sul de Mato Grosso. Boletim Gaùcho de Geografia. No. 32. p. 37-56. Porto Alegre - RS: 2007.

ESSELIN, P. M; OLIVEIRA, T. C. M. Land where cattle created man and defined the latifùndio. History: Debates and Trends. Vol. 7. No. 2. Passo Fundo: 2008.

FADEMS. **Work Plan - Execution of Management Plans for Conservation Units (APAs and Municipal Parks) in the Apa River Basin and Design of the Ecological Broker for the entire Basin.** Fundaçao de Apoio ao Desenvolvimento da Educaçao de Mato Grosso do Sul - UFMS. Ponta Pora, 2012.

FERREIRA, C. C. **Geotechnologies Applied to the Creation and Organization of a Geoenvironmental Database for the Sucuriù River Basin - MS/BR.** 2011. 193 f. Dissertation (Master's Degree) - Federal University of Mato Grosso do Sul, Três Lagoas, 2011.

GONZALEZ, I. B; FREIRE, C. F; MORENTE, L. M. F; ASENSIO, E. P. ***Geographic Information Systems and Research in Human and Social Sciences.*** *CECEL-CSIC. Madrid,* 2012.

GOROSTIAGA, O. L, GONZALEZ, E., DE LLAMAS, P., MOLINAS, A., FRANCO, E., GARCIA, S., Y RIOS, E. **Study of Soil Reconnaissance, Land Use Capacity and Preliminary Land Use Planning Proposal for the Eastern Region of Paraguay. Land use rationalization project.** SSERNMA/MAG/World Bank. Asunción, Paraguay, 1995.

IBGE. **Land Use Technical Manual**. Rio de Janeiro: IBGE, 2006. 91 p. (Technical manuals in Geosciences, n. 7).

LANDIM, P. M. B. **Introduction to spatial estimation methods for map making**. DGA, IGCE, UNESP/Rio Claro, Lab. Geomatemàtica, Texto Didâtico 02, 20 pp. 2000. Available at:

<htpp://www.rc.unesp.br/igce/aplicada/textodi.html>

LANDIS, J. R.; KOCH, G. G. **The measurement of observer agreement for categorical data**. Biometrics, v.33, n.1, 1977. p. 159-174.

LANG, S; BLASCHE, T. **Landscape Analysis with GIS**. Translated by Hermann Kux. Sao Paulo: Oficina de Textos, 2009.

LEPSCH, I. F. **Formação e Conservaçâo dos Solos.** Sao Paulo: Oficina de Textos,

2002.

LOMBARDO, M. A. **The climate and the city**. Climatological Bulletin. Presidente Prudente: ed.UNESP. Year 1, n.2, 1996.

LÓPEZ, J. F. Aguas internacionales. **Manual de ciudadania global**. Mexico, DF: UNEP (United Nations Environment Program), 2004.

MARTINELLI, M. **Mapas de Geografia e Cartografia Temàtica**. Sao Paulo: Contexto, 2003. 112p.

MATO GROSSO DO SUL. **Geoenvironmental Macrozoning of the State of Mato Grosso do Sul.** Campo Grande, 1985.

MENDONÇA, F. **Diagnosis and environmental analysis of a watershed - Methodological proposal from the perspective of zoning, planning and environmental management**. RA'EGA - Geographical space in analysis. Paranà, No. 3. Year III, 1999.

MOLINAS, A. S. M. ***Importancia, Potencialidades y Desafios del Sector AgroRural Paraguayo***. Asunción, 2009. Available at:

<http://www.mag.gov.py/varias/SECTOR%20AGRORUAL%20PYO%202009%20Alfr edo%20Molinas.pdf>

MOREIRA, M. A. **Fundamentals of remote sensing and application methodologies**. 2. ed. Viçosa: UFV, 2005. 307 p.

MOREIRA. C. V. R.; NETO A. G. P. **Climate and Relief**. IN: OLIVEIRA, A. M.; BRITO, S. N. Geologia de Engenharia, Associaçao Brasileira de Geologia de Engenharia. 1st ed. Sao Paulo, 1998, p.69-85.

NEVES, L. P. **Adoption of the Party in Architecture**. Salvador: ed. UFBA, 1989.

OLIVEIRA. M. C. de. **Construction of a chart for determining slope orientation.** Journal of the Geological Institute. Sao Paulo. Jan/Dec. 1984. 47-50p.

PARAGUAY. ***Forestry Law No. 422****. Asuncion: 1973.* Available at: http://www.seam.gov.py/legislaciones-ambientales.html. Accessed on: February 18, 2013.

PARAGUAY. ***Geological Map of Paraguay****. Asuncion: 1986.*

PARAGUAY. ***Law No. 3.239 on Water Resources in Paraguay****. Asuncion: 2007.* Available at: http://www.seam.gov.py/legislaciones-ambientales.html. Accessed on: February 18, 2013.

PARAGUAY. Risk Management Unit (UGR), Ministry of Agriculture and Livestock (MAG). ***BOLETIN "Unidad de Gestión de Riesgos".*** *Asunción, 2007*

PEREIRA, N. R. et al. **Low intensity reconnaissance survey of the soils of the municipality of Bonito, state of Mato Grosso do Sul.** Electronic data. Rio de Janeiro: Embrapa Solos, 2007.

PEREIRA, M. C. B; BROCH, S. O; MEDEIROS, Y. **One river, two countries**. In: BROCH, S. O; MEDEIROS, Y; SOUZA, P. R (Org.). Pé na àgua: uma abordagem transfronteiriça da bacia do Apa. 128 p. Ed. UFMS. Campo Grande - MS: 2008.

PISSINATI, M. C; ARCHELA, R. S. **Geosistema Territòrio e Paisagem - Método de estudo da paisagem rural sob a òtica Bertrandiana**. Revista Gegrafia. V. 18. n. 1. Universidade Estadual de Londrina-Departamento de Geociências. jan/jun. 2009. Available at: < http://www.uel.br/revistas/uel/index. php/geografia/>.

Brazilian Basic Geological Survey Program. Campo Grande, Sheet SF. 21. Scale 1:1,000,000. State of Mato Grosso do Sul. Organized by Hélio de Oliveira Godoi and Edson Gaspar Martins. Brasilia: **CPRM, 1999.**

POTT, A.; POTT, V. J. **Native plants for recovering degraded areas and replenishing vegetation in Mato Grosso do Sul.** Campo Grande: Embrapa Gado de Corte, 2003 (Embrapa Gado de Corte. Comunicado Técnico 75).

RODRIGUEZ, J. M. M; SILVA, E. V; LEAL, A. C. **Environmental Planning in Watersheds.** In: SILVA, E. V; RODRIGUEZ, J. M. M; MEIRELES, A. J. de A. (Org.). Planejamento Ambiental e Bacias Hidrogràficas. Fortaleza: Edições UFC, 2011.

ROSA, R; LEITE, M. E. **Geography and Geotechnologies in Urban Studies.** Caminhos de Geografia - On-line journal. p. 180-186. February, 2006. Available at: http://www.ig.ufu.br/revista/caminhos.html.

ROSA, R. **Geotecnologias na geografia aplicada**. Journal of the Department of Geography. Sao Paulo. v.16. p. 81-90, 2005.

ROSS. J. L. S. **O Registro Cartogràfico dos Fatos Geomórficos e a Questao da Taxonomia do Relvo**. Revista do Departamento de Geografia, Sao Paulo, n. 8, p. 63-74, 1992.

ROSS, J. L. S. **Empirical Analysis of the Fragility of Natural and Anthropized Environments.** Revista do Departamento de Geografia. n.8, p.63-74, 1994.

SALGUEIRO, T. B. **Landscape and Geography.** Finisterra, XXXVI, 72, p. 37-53, 2001.

STEINKE, V. A.; SAITO, C. H. Geoenvironmental assessment of Brazilian territory in transboundary river basins. **Revista Brasileira de Gestão e Desenvolvimento Regional**, Taubaté, SP, v. 6, n. 1, p.189-222, jan-abr/2010.

SANTOS, M. **The Nature of Space: Technique and Time, Reason and Emotion**. 4. ed.

2. reprint. Editora Universidade de Sao Paulo. Sao Paulo, 2006.

SANTOS, M. **Metamorfoses do Espaço Habitado, fundamentos teórico e metodológico da geografia**. Hucitec. Sao Paulo 1988.

State Secretariat for the Environment, Planning, Science and Technology and Mato Grosso do Sul Environment Institute. **Mato Grosso do Sul State Water Resources Plan (PERH-MS).** Campo Grande, MS: Editora UEMS, 2010.

Secretariat for Strategic Affairs (SAE). Water and Sustainable Development Brazil's Border and Transboundary Water Resources. Strategic Studies Series - Water and Sustainable Development. Brasilia: 2013.

SILVA, A. de B. **Geo-referenced Information Systems: concepts and foundations**. Campinas, SP: Editora da Unicamp, 2003.

SILVA, V. R. **Socio-environmental analysis of the Biguaçu-SC river basin: Subsidies for territorial planning and organization.** 2007. 227 f. Thesis (Doctorate) - Faculty of Philosophy, Letters and Human Sciences, University of Sao Paulo, Sao Paulo, 2007.

SOCHAVA, V. B. **The study of geosystems. Métodos em questào**, n.16, IGUSP. Sao Paulo, 1977. 51 p.

SOUZA, M. L. de. **Changing the city: a critical introduction to urban planning and management**. 3rd ed. Bertrand. Rio de Janeiro: 2004.

TRICART, J. **Ecodynamics.** Rio de Janeiro: IBGE, 1977.

TUCCI, C.E.M. (Organizer). Hidrologia: Ciência e Aplicaçao, 3ª edition, Porto Alegre, **Editora da UFRGS/ABRH,** 2004.

VALERIANO, M. de M. **TOPODATA: guide to using local geomorphometric data**. Sao José dos Campos: INPE, 2008.

MIX
Papier aus verantwortungsvollen Quellen
Paper from responsible sources
FSC® C105338

Printed by Books on Demand GmbH, Norderstedt / Germany